SpringerBriefs in Applied Sciences and Technology

W0037572

More information about this series at http://www.springer.com/series/8884

Sujoy Kumar Saha • Gian Piero Celata

Critical Heat Flux in Flow Boiling in Microchannels

 Springer

Sujoy Kumar Saha
Indian Institute of Engineering Science
  and Technology, Shibpur
Howrah, India

Gian Piero Celata
ENEA Division of Advanced Technologies
  for Energy and Industry
Rome, Italy

ISSN 2191-530X   ISSN 2191-5318   (electronic)
SpringerBriefs in Applied Sciences and Technology
ISBN 978-3-319-17734-2   ISBN 978-3-319-17735-9   (eBook)
DOI 10.1007/978-3-319-17735-9

Library of Congress Control Number: 2015937217

Springer Cham Heidelberg New York Dordrecht London

Printed on acid-free paper

Springer International Publishing AG Switzerland is part of Springer Science+Business Media
(www.springer.com)

Preface

Critical heat flux (CHF) in flow boiling in microchannels has been discussed in this book. The state-of-the-art review has been made. Investigations on critical heat flux in flow boiling in microchannels have been dealt with. Predictions of CHF, models, and correlations have been discussed in detail. This is followed by studies of boiling and CHF and some parametric effects on CHF. The book ends with conclusion and delineation of further research.

Contents

Nomenclature

A_{inner}	Inner area (m^2)
AMP	Oscillation amplitude
Cp	Thermal capacity (J/kg °C)
D	Diameter (mm)
D_{inner}	Inner diameter (cm)
DP_{max}	The maximum instant pressure drop (kPa)
DP_{min}	Minimum instant pressure drop (kPa)
f	Frequency (Hz)
f_L	Friction factor in laminar flow
f_T	Friction factor in turbulent flow
FS	Full scale
G	Mass flux (kg/m^2 s)
I	Current (A)
i_{in}	The liquid enthalpy at the channel inlet (J/kg)
i_l	The enthalpy of saturated liquid at the system pressure (J/kg)
i_{lv}	The latent heat of evaporation at the system pressure (J/kg)
i_{sub}	$i_l - i_{in}$ (J/kg)
i_v	The enthalpy of saturated vapor at the system pressure (J/kg)
L	Tube length (mm)
L(t)	Instant bubble length in slug flow at time t (μm)
$L(t_0)$	Bubble length in slug flow at the reference time, t_0 (μm)
L_{heated}	Heated length (cm)
L_r	Bubble length ratio, $L(t)/L(t_0)$
M	Total number of recorded data points
MAX	Maximum value
N_{pch}	Sub-cooling number, $\left(\dfrac{Q_c v_{lv}}{W i_{lv} v_l} \right)$
N_{sub}	Phase change number, $\left(\dfrac{i_{sub} v_{lv}}{i_{lv} v_l} \right)$

P   Pressure (kPa)
Q'' Heat flux (W/cm^2)
Q_c The heat transfer rate to the channel (J/s)
Re   Reynolds number
T   Temperature (°C)
t   Time (s)
t_0 Reference time (s)
V   Voltage (V)
v_l The specific volume of saturated liquid at the system pressure (m^3/kg)
v_{lv} The specific volume difference between vapor and liquid at the system pressure (m^3/kg)
v_v The specific volume of saturated vapor at the system pressure (m^3/kg)
W   The total mass flow rate to the channels (kg/s)
x_e The exit vapor quality
y   Measured parameter (T or P)

Greek symbol

β Coefficient of exponent for bubble growth in slug flow (s^{-1})

Chapter 1
Introduction

With the advancement of technology, the world is moving towards miniaturization and it is necessary to remove high heat flux. Heat generated per unit area has measured up to 104 W/cm^2 (nuclear reactor). Microchannels and minichannels are naturally well suited for this task, as they provide large heat transfer surface area per unit fluid flow volume [1].

Heat flux removal requirement varies significantly. For densely packed integrated circuit (ICs) [2, 3] and laser mirror [4] the maximum power flux reported is 102 W/cm^2, aviation and VLSI industry need up to 103 W/cm^2 [5] and fusion reactor and defence application often deal with 104 W/cm^2 [6–8]. Heat dissipation requirement will continue to rise with more advancement in technologies and further reduction in the size of these applications. Considering above facts, it can be concluded that microchannel heat sinks seem to be the plausible solution of twenty first century cooling problems.

Tuckerman and Pease [9] had developed microchannel heat sink made up of silicon to remove heat flux of 790 W/cm^2 with water as working fluid. Keyes [10] carried out theoretical analysis of finned microchannel heat sink. Missaggia et al. [11] developed a microchannel heat sink for cooling of two dimensional high power density diode laser arrays.

Flow boiling behavior significantly differs with the channel hydraulic diameter. Macroscale correlation for flow boiling cannot be directly used for microscale, specifically because of the various influences of gravitational force and capillary force. Introduction of flow boiling in micro-channel harnesses the advantage of enhanced single-phase convective flow and high heat transfer rate of boiling.

Understanding of the heat transport helps understanding of the mass and momentum transport inside the microchannel [12–17]. For high heat flux flow boiling, the critical heat flux (CHF) or burnout is very important. Flow boiling heat transfer in micro-channels is an important area of research for removal of high heat flux of the

© Springer International Publishing Switzerland 2015
S.K. Saha, G.P. Celata, *Critical Heat Flux in Flow Boiling in Microchannels*,
SpringerBriefs in Applied Sciences and Technology,
DOI 10.1007/978-3-319-17735-9_1

order of 300 W/cm² in computer chips. Flow boiling cooling is much better than single phase liquid cooling since the former method involves latent heat of vaporization.

Several non-dimensional numbers are used for flow boiling in microchannels [18–24]. Two new nondimensional groups relevant to flow boiling phenomenon were proposed by Kandlikar [13].

Figures 1.1, 1.2, 1.3, 1.4 and 1.5 show the heat transfer mechanisms and systems for flow boiling in microchannels [5]. The data of Huo et al. [25], Shiferaw et al. [26] and Chen et al. [27] show that, data from the same experimental set up may vary for different observations.

$E_o = 1.6$ proposed in [28] is not always the transition criterion from macroscale to microscale. References [29–34] deal with the transition from macroscale to microscale flow boiling heat transfer in detail.

The most important cause of flow boiling is heating, and it is this type of boiling that this book is concerned with. Evaporation in forced convection is due to variations in temperature, velocity, or even contact angle [35]. Correlation of the slip ratio characterizes two-phase hydrodynamics of boiling in a channel. Slip ratio correlations for boiling of water can be obtained from Butterworth [36].

Klimenko [37, 38] and Kandlikar [39] have observed that in the fully developed region of sub cooled boiling, the effect of convective heat transfer is insignificant and the heat transfer is mainly due to nucleate boiling.

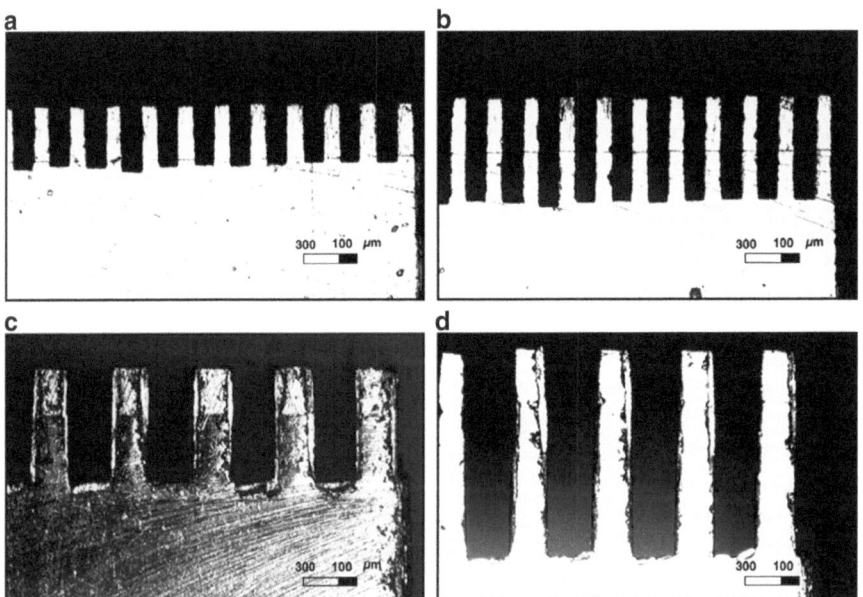

Fig. 1.1 Microscope images of microchannel [5]

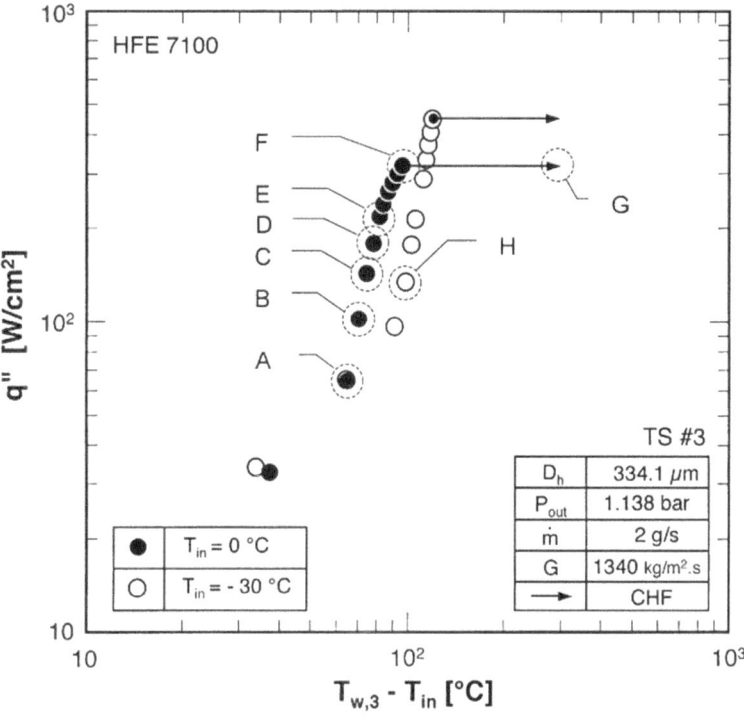

Fig. 1.2 Subcooled boiling curve [5]

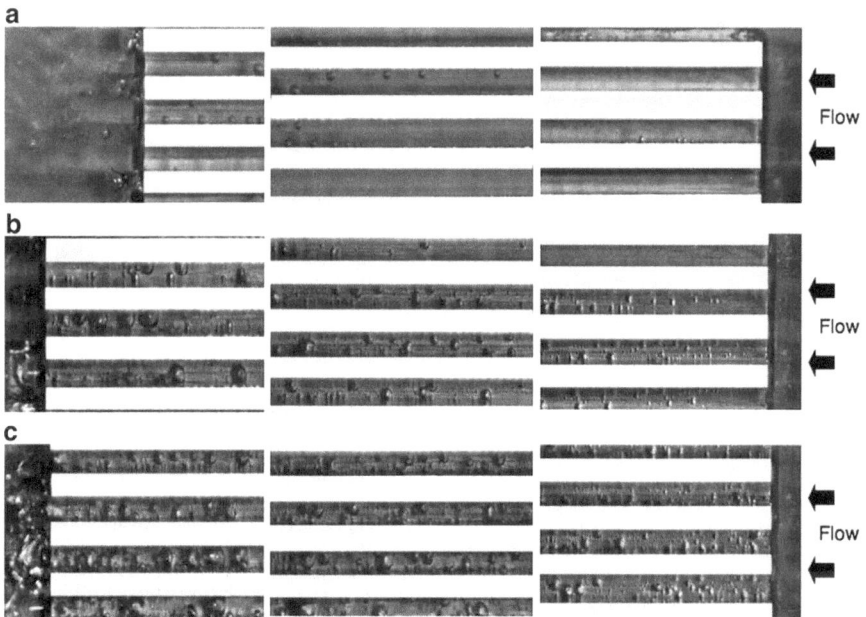

Fig. 1.3 Flow boiling images [5]

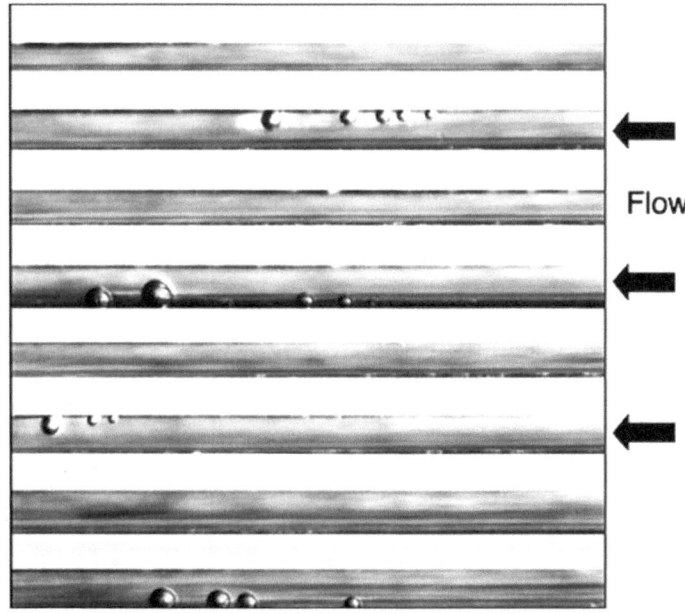

Fig. 1.4   Flow boiling image of middle region of microchannel [5]

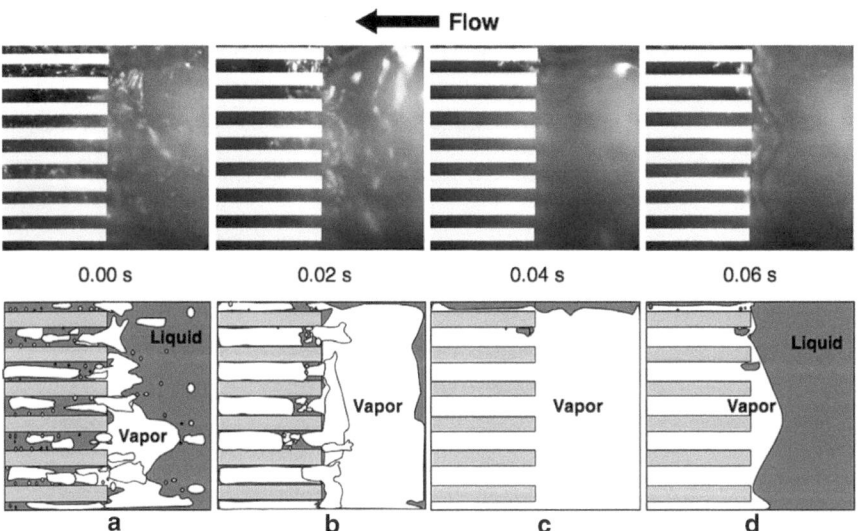

Fig. 1.5   Premature CHF and flow oscillations [5]

The study of different flow regimes is important to predict the pressure drop and heat transfer characteristics. This needs high speed photography. In conventional channels, as explained by Thome [34], the sequence of flow pattern is bubbly, slug, churn, wispy-annular and annular flow in vertical flow, whereas for the horizontal flow, bubbly, slug, plug, annular, stratified, annular with mist and wave flow exist. In case of microchannels, flow patterns are quite different from conventional channels. Simple extrapolation does not work. Pfahler et al. [40] carried out experiments on three different microchannels.

Sobierska et al. [41] performed experiments using water in a single rectangular microchannel and observed bubbly, slug and annular flow. Lee and Mudawar [5] visualized the nucleate flow boiling at inlet, middle and outlet section of microchannel and related the flow patterns with boiling curve. Megahed and Hassan [42] carried out experiments on 45 rectangular microchannels. Zhang and Fu [43] performed experiments on vertically upward microtubes using liquid nitrogen as working fluid. They reported bubbly flow, slug flow, churn flow and annular flow as the main flowpatterns. Two types of flow pattern were reported by Kawahara et al. [44]; (a) quasi-homogeneous (b) quasi-separated. Choi and Kim [45] carried out water and nitrogen-gas two phase flow experiments over a range of superficial velocity with five different types of rectangular microchannels. Choi et al. [46] observed bubbly flow, transition flow and liquid ring flow on rectangular microchannels of varying aspect ratio. Additional references [16, 47–60] deal with the two-phase flow patterns in detail.

Use of flow map for predicting two phase pattern is extremely popular and well established for conventional size channels [61–65]. The most commonly used predictive tools for flow regimes (patterns) are flow regime maps. Perhaps, Triplett et al. [66] were the first to develop flow map for microchannel. Harirchian and Garimella [67] developed comprehensive flow regime map for FC-77. References [68–70] deal with flow pattern maps in detail. References [71–82] deal with the bubble dynamics. References [25, 32, 37–39, 83–140] deal with heat transfer models and correlations for flow boiling in microchannels.

References

1. Mudawar I (2011) Two-phase microchannel heat sink theory, application and limitations. J Electron Packag 133(4):041002
2. Ali R (2010) Phase change phenomena during fluid flow in micro channels. Doctoral thesis, Royal Institute of Technology, Stockholm
3. Pop E, Goodson KE (2006) Thermal phenomena in nanoscale transistors. J Electron Packag 128:102–108
4. Phillips RJ (1988) Microchannel heat sinks. MIT Lincoln Lab J 1:31–47
5. Lee J, Mudawar I (2008) Fluid flow and heat transfer characteristics of low temperature two-phase micro-channel heat sinks—part 1: experimental methods and flow visualization results. Int J Heat Mass Transf 51:4315–4326

6. Boyd RD (1985) Subcooled flow boiling critical heat flux and its application to fusion energy components—part 1: a review of fundamentals of CHF and related data base. Fusion Technol 7:7–30
7. Lee J, Mudawar I (2009) Low-temperature two-phase microchannel cooling for high heat-flux thermal management of defense electronics. IEEE Trans Compon Packag Technol 32:453–465
8. Kandlikar SG (2005) High flux heat removal with microchannels—a roadmap of challenges and opportunities. Heat Transf Eng 26:5–14
9. Tuckerman DB, Pease RFW (1981) High-performance heat sinking for VLSI. IEEE Electron Device Lett 2(198):126–129
10. Keyes RW (1984) Heat transfer in forced convection through fins. IEEE Trans Electron Devices 311:1218–1221
11. Missaggia IJ, Walpole JN, Liau ZL, Phillips RJ (1989) Microchannel heat sinks for two-dimensional high-power-density diode laser arrays. IEEE J Quantum Electron 25:1988–1992
12. Incropera FP, Bergman TL, Lavine AS (2007) Fundamentals of heat and mass transfer, 6th edn. Wiley, New York
13. Kandlikar SG (2004) Heat transfer mechanisms during flow boiling in microchannels. J Heat Transf 126:8–16
14. Kandlikar SG, Grande WJ (2002) Evolution of microchannel flow passage thermohydraulic performance and fabrication technology. In: ASME international mechanical engineering congress and exposition, New Orleans, Louisiana, 17–22 November
15. Mehendale SS, Jacobi AM, Shah RK (2000) Fluid flow and heat transfer at micro and meso-scales with applications to heat exchanger design. Appl Mech Rev 53:175–193
16. Cornwell K, Kew PA (1993) Boiling in small parallel channels. In: Pilavachi PA (ed) Energy efficiency in process technology. Elsevier, New York, pp 624–638 (Chap. 7)
17. Kew P, Cornwell K (1997) Correlation for prediction of boiling heat transfer in small diameter channel. J Therm Eng 17:705–715
18. Steinke ME, Kandlikar SG (2003) Flow boiling and pressure drop in parallel microchannels. In: Proceedings of first international conference on microchannels and minichannels, Rochester, New York, 24–25 April, pp 567–579
19. Yan Y, Lin T (1998) Evaporation heat transfer and pressure drop of refrigerant R-134a in a small pipe. Int J Heat Mass Transf 41:4183–4194
20. Wambsganss MW, France DM, Jendrzejczyk JA, Tran TN (1993) Boiling heat transfer in horizontal small-diameter tube. J Heat Transf 115:963–972
21. Lin S, Kew PA, Cornwell K (2001) Flow boiling of refrigerant R141B in small tubes. Trans IChemE 79(A):417–424
22. Yen TH, Kasagi N, Suzuki Y (2002) Forced convective boiling heat transfer at low mass and heat fluxes. In: Proceedings of the international symposium on compact heat exchangers, Grenoble, Edizioni ETS, p 190
23. Kamidis DE, Ravigururajan TS (1999) Single and two-phase refrigerant flow in minichannels, In: Proceedings of 33rd national heat transfer conference, Albuquerque, New Mexico, 15–17 August
24. Baldassari C, Marengo M (2013) Flow boiling in microchannels and microgravity. Prog Energy Combust Sci 39:1–36
25. Huo X, Chen L, Tian YS, Karayiannis TG (2004) Flow boiling and flow regimes in small diameter tubes. Appl Therm Eng 24:1225–1239
26. Shiferaw D, Huo X, Karayiannis TG, Kenning DBR (2007) Examination of heat transfer correlations and a model for flow boiling of R134a in small diameter tubes. Int J Heat Mass Transf 50:5177–5193
27. Chen L, Tian YS, Karayiannis TG (2006) The effect of tube diameter on vertical two phase flow regimes in small tubes. Int J Heat Mass Transf 49:4220–4230
28. Ullmann A, Brauner N (2007) The prediction of flow pattern maps in minichannels. Multiph Sci Technol 19(1):49–73

29. Morini GL (2008) Single-phase convective heat transfer in laminar and transitional regime in microchannels. In: ECI international conference on heat transfer and fluid flow in microscale, Whistler, 21–26 September
30. Celata GP, Lorenzini M, McPhail SJ, Morini GL, Zummo G (2008) Experimental analysis of liquid forced convection in rough microtubes. In: Proceedings of 5th eurotherm European thermal science conference, Eindhoven, 19–22 May
31. Kandlikar SG (2001) Two-phase flow patterns, pressure drop and heat transfer during boiling in minichannel and microchannel flow passages of compact heat exchanger. In: Compact heat exchangers and enhancement technology for the process industries, Begell House, New York, pp 319–334
32. Thome JR (2004) Boiling in microchannels: a review of experiment and theory. Int J Heat Fluid Flow 25:128–139
33. Thome JR, Dupont V, Jacobi AM (2004) Heat transfer model for evaporation in microchannels, part I: presentation of the model. Int J Heat Mass Transf 47:3375–3385
34. Thome JR (2004) Engineering data book III. Wolverine Tube, Inc., Huntsville (Chap. 12)
35. Som T, Kumar S, Kulkarni VN (1999) H+ ion induced hydrogen depletion from a-C: H films. Nucl Instr Meth Phys Res B 156:212–216
36. Butterworth DA (1975) A comparison of some void-fraction relationships for co-current gas-liquid flow. Int J Multiph Flow 1:845–850
37. Klimenko VV (1988) A generalized correlation for two phase forced flow heat transfer. Int J Heat Mass Transf 31:541–552
38. Klimenko VV (1990) A generalized correlation for two phase flow heat transfer, second assessment. Int J Heat Mass Transf 33:2073–2088
39. Kandlikar SG (1998) Heat transfer characteristics in partial boiling, fully developed boiling, and significant void flow regions of sub cooled flow boiling. J Heat Transf 120:395
40. Pfahler J, Harley J, Bau H, Zemel J (1990) Liquid transport in micron and sub micron channels. Sens Actuators A 22:431–434
41. Sobierska E, Kulenovic R, Mertz R, Groll M (2006) Experimental results of flow boiling of water in a vertical microchannel. Exp Therm Fluid Sci 31:111–119
42. Megahed A, Hassan I (2009) Two-phase pressure drop and flow visualization of FC-72 in a silicon microchannel heat sink. Int J Heat Fluid Flow 30:1171–1182
43. Zhang P, Fu X (2009) Two-phase flow characteristics of liquid nitrogen in vertically upward 0.5 and 1.0 mm micro-tubes: visualization studies. Cryogenics 49:565–575
44. Kawahara A, Sadatomi M, Nei K, Matsuo H (2009) Experimental study on bubble velocity, void fraction and pressure drop for gas-liquid two-phase flow in a circular microchannel. Int J Heat Fluid Flow 30:831–841
45. Choi C, Kim M (2011) Flow pattern based correlations of two-phase pressure drop in rectangular microchannels. Int J Heat Fluid Flow 32:1199–1207
46. Choi CW, Yu DI, Kim MH (2011) Adiabatic two-phase flow in rectangular microchannels with different aspect ratios: part I—flow pattern, pressure drop and void fraction. Int J Heat Mass Transf 54:616–624
47. Kasza KE, Didascalou T, Wambsganss MW (1997) Microscale flow visualization of nucleate boiling in small channels: mechanisms influencing heat transfer. In: Proceedings of the international conference on compact heat exchanges for the process industries, Begell House Inc., New York, pp 343–352
48. Chung PMY, Kawaji M (2004) The effect of channel diameter on adiabatic two phase flow characteristics in microchannels. Int J Multiph Flow 30:735–761
49. Lee J, Mudawar I (2005) Two-phase flow in high-heat-flux micro-channel heat sink for refrigeration cooling applications: part II. Heat transfer characteristics. Int J Heat Mass Transf 48:941–955
50. Chen T, Garimella SV (2006) Measurements and high-speed visualizations of flow boiling of a dielectric fluid in a silicon microchannel heat sink. Int J Multiph Flow 32:957–971
51. Wang G, Cheng P, Bergles AE (2008) Effects of inlet/outlet configurations on flow boiling instability in parallel microchannels. Int J Heat Mass Transf 51:2267–2281

52. Kashid MN, Gerlach I, Goetz S, Franzke J, Acker JF, Platte F, Agar DW, Turek S (2005) Internal circulation within the liquid slugs of liquid-liquid slug flow capillary micro-reactor. Ind Eng Chem Res 44:5003–5010
53. Agostini B, Revellin R, Thome JR (2008) Elongated bubbles in microchannels. Part I: experimental study and modeling of elongated bubble velocity. Int J Multiph Flow 34:590–601
54. Schilder B, Man SYC, Kasagi N, Hardt S, Stephan P (2011) Flow visualization and local measurement of forced convection heat transfer in a microtube. J Heat Transf 132(3):031702
55. Lee JY, Kim MH, Kaviany M, Son SY (2011) Bubble nucleation in microchannel flow boiling using single artificial cavity. Int J Heat Mass Transf 54:5139–5148
56. Krishnamurthy S, Peles Y (2010) Flow boiling heat transfer on micro pin fins entrenched in a microchannel. J Heat Transf 132(4):041007
57. Barber J, Brutin D, Sefiane K, Tadrist L (2010) Bubble confinement in flow boiling of FC-72 in a "rectangular" microchannel of high aspect ratio. Exp Therm Fluid Sci 34:1375–1388
58. Megahed A (2011) Experimental investigation flow boiling characteristics in a cross-linked microchannel heat sink. Int J Multiph Flow 37:380–393
59. Edel ZJ, Mukherjee A (2011) Experimental investigation of vapor bubble growth during flow boiling in a microchannel. Int J Multiph Flow 37:1257–1265
60. David MP, Steinbrenner JE, Miler J, Goodson KE (2011) Adiabatic and diabatic two-phase venting flow in a microchannel. Int J Multiph Flow 37:1135–1146
61. Hewitt GF, Roberts DN (1969) Studies of two-phase flow patterns by simultaneous x-ray and flash photography. Atomic Energy Research Establishment, Harwell, report no. AERE-M 2159
62. Taitel Y, Dukler AE (1976) A model for predicting flow regime transitions in horizontal and near horizontal gas-liquid flow. AIChE J 22:47–55
63. Kattan N, Thome JR, Favrat D (1998) Flow boiling in horizontal tubes. Part I: development of a diabatic two-phase flow pattern map. J Heat Transf 120:140–147
64. Kattan N, Thome JR, Favrat D (1998) Flow boiling in horizontal tubes: part II: new heat transfer data for five refrigerants. J Heat Transf 120:148–155
65. Kattan N, Thome JR, Favrat D (1998) Flow boiling in horizontal tubes: part III: development of a new heat transfer model based on flow patterns. J Heat Transf 120:156–165
66. Triplett KA, Ghiaasiaan SM, Abdel-Khalik SI, Sadowski DL (1999) Gas-liquid two-phase flow in microchannels, part I: two-phase flow patterns. Int J Multiph Flow 25:377–394
67. Harirchian T, Garimella SV (2010) A comprehensive flow regime map for microchannel flow boiling with quantitative transition criteria. Int J Heat Mass Transf 53:2694–2702
68. Sur A, Liu D (2012) Adiabatic air-water two-phase flow in circular microchannels. Int J Therm Sci 53:18–34
69. Hewitt GF, Robert DN, Studies in two phase flow patterns by simultaneous x-ray and flash photography. AERE-M-2159 HMSO
70. Baker D (1954) Simultaneous flow of oil and gas. Oil Gas J 53:183–195
71. Lee PC, Tseng FG, Pan C (2004) Bubble dynamics in microchannels, part I: single microchannel. Int J Heat Mass Transf 47:5575–5589
72. Li HY, Tseng FG, Pan C (2004) Bubble dynamics in microchannels, part II: two parallel microchannels. Int J Heat Mass Transf 47:5591–5601
73. Liu D, Lee PS, Garimella SV (2005) Prediction of the onset of nucleate boiling in microchannels flow. Int J Heat Mass Transf 48:5134–5149
74. Hsu YY (1962) On the size range of active nucleation cavities on a heating surface. J Heat Transf 84:207–216
75. Meder S (2007) Study on bubble growth rate in a single microchannel heat exchanger with high-speed CMOS-camera. Master thesis, Swiss Federal Institute of Technology, Zurich and Stanford University California
76. Bogojevic D, Sefiane K, Duursma D, Walton AJ (2013) Bubble dynamics and flow boiling instabilities in microchannels. Int J Heat Mass Transf 58:663–675

77. Lee M, Wong YY, Wong M, Zohar Y (2003) Size and shape effects on two-phase flow patterns in microchannel forced convection boiling. J Micromech Microeng 13:155–164
78. Fu X, Zhang P, Huang CJ, Wang RZ (2010) Bubble growth, departure and the following flow pattern evolution during flow boiling in a mini-tube. Int J Heat Mass Transf 53:4819–4831
79. Gedupudi S, Zu YQ, Karayiannis TG, Kenning DBR, Yan YY (2011) Confined bubble growth during flow boiling in a mini/micro-channel of rectangular cross-section part I: experiments and 1-D modeling. Int J Therm Sci 50:250–266
80. Yin L, Jia L, Guan P, Liu F (2012) An experimental investigation on the confined and elongated bubbles in subcooled flow boiling in a single microchannel. J Therm Sci 21:549–556
81. Tuo H, Hrnjak P (2014) Visualization and measurement of periodic reverse flow and boiling fluctuations in a microchannel evaporator of an air-conditioning system. Int J Heat Mass Transf 71:639–652
82. Collier JG, Thome JR (1996) Convective boiling and condensation, 3rd edn. Oxford University, New York
83. Stephan P, Hammer J (1994) A new model for nucleate boiling heat transfer. Int J Heat Mass Transf 30:119–125
84. Rohsenow WM (1952) A method of correlating heat transfer data for surface boiling of liquids. Trans AMSE 74:969–975
85. Stephan K, Abdelsalam M (1980) Heat transfers correlations for natural convection boiling. Int J Heat Mass Transf 23:73–87
86. Cooper MG (1984) Saturated nucleate pool boiling—a simple correlation. In: Proceedings of the 1st UK national heat transfer conference, IChemE symposium series, vol 86, pp 785–793
87. Tong LS, Tang YS (1997) Boiling heat transfer and two-phase flow, 2nd edn. Taylor & Francis, Bristol
88. Carey VP (1992) Liquid phase change phenomena: an introduction to the thermo physics of vaporization and condensation process in heat transfer equipment. Hemisphere, New York
89. Gungor KE, Winterton RHS (1987) Simplified general correlation for saturated flow boiling and comparisons of correlations with data. Chem Eng Res Des 65:148–156
90. Chen JC (1966) Correlation for boiling heat transfer to saturated fluids in convective flow. Ind Eng Chem Process Des Dev 5:322–329
91. Shah MM (1977) A general correlation for heat transfer during subcooled boiling in pipes and annuli. ASHRAE Trans 83:202–215
92. Dittus FW, Boelter LMK (1930) Heat transfer in automobile radiators of tubular type. Univ Calif Pub Eng 2:443–461
93. Bjorge RW, Hall GR, Rohsenow WM (1982) Correlations of forced convective boiling heat transfer data. Int J Heat Mass Transf 25:753–757
94. Liu ZL, Winterton RHS (1991) A general correlation for saturated and sub-cooled flow boiling in tubes and annuli, based on a nucleate pool boiling equation. Int J Heat Mass Transf 34(11):2759–2766
95. Kandlikar SG (1990) A general correlation for saturated two-phase flow boiling heat transfer inside horizontal and vertical tubes. J Heat Transf 112:219–228
96. Kandlikar SG (1991) Development of a flow boiling map for sub-cooled and saturated flow boiling of different fluids inside circular tubes. J Heat Transf 113:190–200
97. Gnielinski V (1970) New equations for heat and mass transfer in turbulent pipe and channel flow. Int Chem Energy 16:359–368
98. Petukov BS (1970) Heat transfer and friction in turbulent pipe flow with variable physical properties. Adv Heat Transf 6:503–565
99. Forster HK, Zuber N (1955) Bubble dynamics and boiling in heat transfer. AIChE J 1:532–535, 1955154
100. Steiner D, Taborek J (1992) Flow boiling heat transfer in vertical tubes correlated by asymptotic model. Heat Transf Eng 13:43–69
101. Thome JR Laboratory of heat and mass transfer. Faculty of Engineering Science, Swiss Federal Institute of Technology, Lausanne, Switzerland

102. Agostini B, Watel B, Bontemps A, Thonon B (2004) Liquid flow friction factor and heat transfer coefficients in small channels: an experimental investigation. Exp Therm Fluid Sci 28:97–103

103. Cooper MG (1989) Flow boiling—the "apparently nucleate" regime. Int J Heat Mass Transf 32:459–464

104. Kenning DBR, Cooper MG (1989) Saturated flow boiling of water in vertical tubes. Int J Heat Mass Transf 32:445–458

105. Jung DS, McLinden M, Radermacher R, Didion D (1989) A study of flow boiling heat transfer with refrigerant mixtures. Int J Heat Mass Transf 32:1751–1764

106. Jung DS, Radermacher R (1991) Prediction of heat transfer coefficients of various refrigerants during evaporation. ASHRAE Trans 97(2):48–53

107. Webb RL, Gupte NS (1992) A critical review of correlations for convective vaporization in tubes and tube banks. Heat Transf Eng 13:58–81

108. Chen JC (1966) Correlation for boiling heat transfer to saturated fluids in convective flow. AIChE Proc Des Dev 5(3):322–329

109. Forster HK, Zuber N (1955) Dynamics of vapor bubble growth and boiling heat transfer. AIChE J 1:531–535

110. Dittus EJ, Boelter LMK (1930) Heat transfer in automobile radiators of tubular type. Univ Calif Berkley Publ Eng 2:443–461

111. Bennet DL, Chen JC (1980) Forced convective boiling in vertical tubes for saturated pure components and binary mixtures. AIChE J 26:454–461

112. Gungor KE, Winterton RHS (1986) A general correlation for flow boiling in tubes and annuli. Int J Heat Mass Transf 29:351–358

113. Prodanovic V, Fraser D, Salcudean M (2002) On the transition from partial to fully developed sub-cooled flow boiling. Int J Heat Mass Transf 45:4227–4738

114. Vlassie C, Macchi H, Guilpart J, Agostini B (2002) Flow boiling in small diameter channels. Int J Refrig 27:191–201

115. Tran TN, Wambsganss MW, France DM (1996) Small circular and rectangular-channel boiling with two refrigerants. Int J Multiph Flow 22(3):485–498

116. Lin S, Kew PA, Cornwell K (1998) Two phase flow regimes and heat transfer in small tubes and channels. In: Proceedings of 11th international heat transfer conference, vol 2, 23–28 August

117. Agostini B, Watel B, Bontemps A, Thonon B (2002) Friction factor and heat transfer coefficient of R134a liquid flow in mini-channels. Appl Therm Eng 22:1821–1834

118. Ghiaasiaan SM, Abdul-Khalik SI (2001) Two phase flow in micro-channels. Adv Heat Transf 34:145–254

119. Lee J, Mudawar I (2004) Two phase flow in high-heat micro-channel heat sink for refrigeration cooling applications: part II: heat transfer characteristics. Int J Heat Mass Transf 48:941–955

120. Lazarek GM, Black SH (1982) Evaporative heat transfer pressure drop and critical heat flux in a small vertical tube with R113. Int J Heat Mass Transf 25(7):945–960

121. Bao ZY, Fletcher DF, Haynes BS (2000) Flow boiling heat transfer of Freon R11 and HCFC123 in narrow passages. Int J Heat Mass Transf 43:3347–3358

122. Mehendale SS, Jacobi AM (2000) Evaporative heat transfer in meso-scale heat exchanger. ASHRAE Trans 106(1):446–452

123. Yu W, France DM, Wambsganss MW, Hull JR (2002) Two-phase pressure drop boiling heat transfer and critical heat flux to water in a small-diameter horizontal tube. Int J Multiph Flow 28:927–941

124. Kew PA, Cornwell K (1997) Correlations for the prediction of boiling heat transfer in small-diameter channels. App Therm Eng 17:705–715

125. Ravigururajan TS (1998) Impact of channel geometry on two-phase flow heat transfer characteristics of refrigerants in micro-channels heat exchangers. J Heat Transf 120:485–491

126. Lee HJ, Lee SY (2001) Heat transfer correlation for boiling flows in small rectangular horizontal channels with low aspect ratios. Int J Multiph Flow 27:2043–2062

127. Lin S, Kew PA, Cornwell K (2001) Two-phase heat transfer to a refrigerant in a 1 mm diameter tube. Int J Refrig 24:51–56

128. Warrier GR, Dhir VK, Momoda LA (2002) Heat transfer and pressure drop in narrow rectangular channels. Exp Therm Fluid Sci 26:53–64
129. Wen DS, Yan Y, Kenning DBR (2004) Saturated flow boiling of water in a narrow channel: time-averaged heat transfer coefficient and correlations. Appl Therm Eng 24:1207–1223
130. Ingo H, Boye H, Jurgen S (2000) Onset of nucleate boiling in mini-channels. Int J Therm Sci 39:505–513
131. Kennedy JE, Roach GM, Dowling MF, Adel-Khalik SI, Ghiaasiaan SM, Jeter SM, Quershi ZH (2000) The onset of flow instability in uniformly heated horizontal micro-channels. Trans ASME 122:118–125
132. Chedester RC, Ghiaasiaan SM (2002) A proposed mechanism for hydro-dynamically controlled onset of significant void in micro-tubes. Int J Heat Fluid Flow 23:769–775
133. Davis EJ, Anderson GH (1966) The incipience of nucleate boiling in forced convection flow. AIChE J 12:774–780
134. Haynes BS, Fletcher DF (2003) Sub cooled flow boiling heat transfer in narrow passages. Int J Heat Mass Transf 46:3673–3882
135. Gorenflo D (1993) Pool boiling. In: VDI Heat Atlas (ed) Chapter Ha. VDI Vela, Dusseldof
136. Koo J-M, Jiang L, Zhang L, Zhou P, Banerjee SS, Kenny TW, Santiago JG, Goodson KE (2001) Modeling of two-phase micro-channel heat sinks for VLSI chips. In: The 14th IEEE international conference on micro electro mechanical systems. doi:10.1109/MEMSYS.906568
137. Zhang L, Wang EN, Koo J (2002) Enhanced nucleate boiling in micro-channels. In: IEEE international conference on micro electro mechanical systems, Las Vegas, pp 89–92
138. Yen T, Kasagi N, Suzuki Y (2003) Forced convective boiling heat transfer in micro-tubes at low mass and heat fluxes. Int J Multiph Flow 29:1771–1792
139. Sumith B, Kaminaga F, Matsumura K (2003) Saturated flow boiling of water in a vertical small diameter tube. Exp Therm Fluid Sci 27:789–801
140. Roach GM, Abdel-Khalik SI, Ghiaasiaan SM, Jeter SM (1999) Low-flow onset of flow instability in heated micro-channels. Nucl Sci Eng 133(1):106

Chapter 2
Critical Heat Flux

2.1 Critical Heat Flux in Flow Boiling in Microchannels

The critical heat flux (CHF) condition is characterized by a sharp reduction of the local heat transfer coefficient as a result of the replacement of liquid by vapor adjacent to the heat transfer surface [1]. The CHF condition in flow boiling can be of different nature [1–5]. At low vapor quality, it is associated with subcooled boiling or saturated boiling and high heat. However, at medium or high quality, it is the dryout and there is no liquid film on the tube wall. Usually this is in case of annular flow and due to surface wave instabilities or entrainment and vaporization.

Critical heat flux can be understood by fluid dynamics, thermodynamics and heat transfer. It is influenced by a large number of process and system variables [6–8]. The scenario during CHF has been studied in [4]. Understanding of CHF mechanisms needs hydrodynamic instability theory [4]. Figure 2.1 [9] shows the variations of the evaporation heat transfer coefficient with the imposed wall heat flux at different mean vapor quality.

Near wall fluid behaves as if a growing bubble is fed by the vapor evaporated from a thin microlayer at its bottom and CHF occurs due to the dryout of the microlayer [10]. Effects of the refrigerant saturated temperature on the evaporation heat transfer coefficient at two different wall heat fluxes are shown in Fig. 2.2 [9].

There is a thermal boundary layer adjacent to the heater surface during nucleate and transition boiling. Depending on the wettability and the temperature of the surface, the vapor stems undergo changes and the merging of the vapor stems triggers CHF [11, 12].

At CHF, there is instability of velocity boundary layer and dryout of microlayer adjacent to the surface seem. Bergles [13] and Kandlikar [14] studied the effects of contact angle, surface orientation and subcooling.

In the microlayer evaporation based model for CHF [15, 16], it is assumed that a dry spot is formed when there are a critical number of bubbles surrounding a single bubble and the liquid supply to the microlayer of the central bubble is restricted.

© Springer International Publishing Switzerland 2015
S.K. Saha, G.P. Celata, *Critical Heat Flux in Flow Boiling in Microchannels*,
SpringerBriefs in Applied Sciences and Technology,
DOI 10.1007/978-3-319-17735-9_2

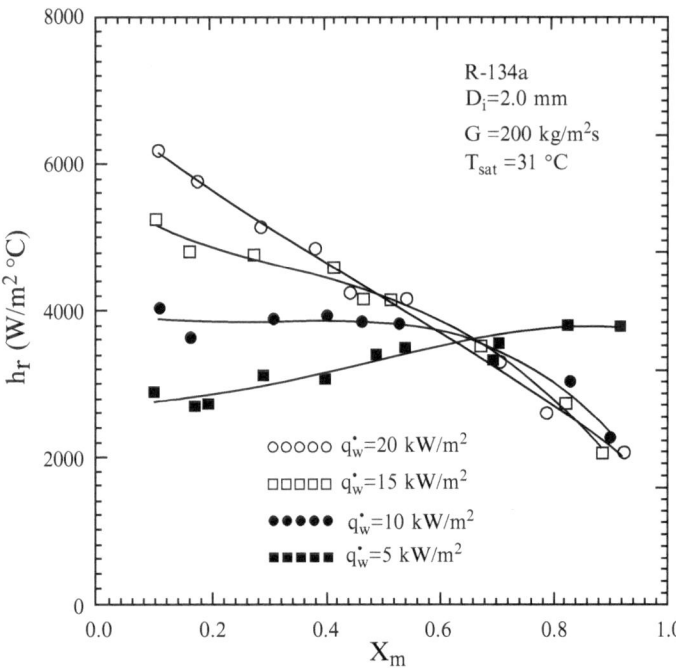

Fig. 2.1 Variations of the evaporation heat transfer coefficient with the imposed wall heat flux at different mean vapor quality [9]

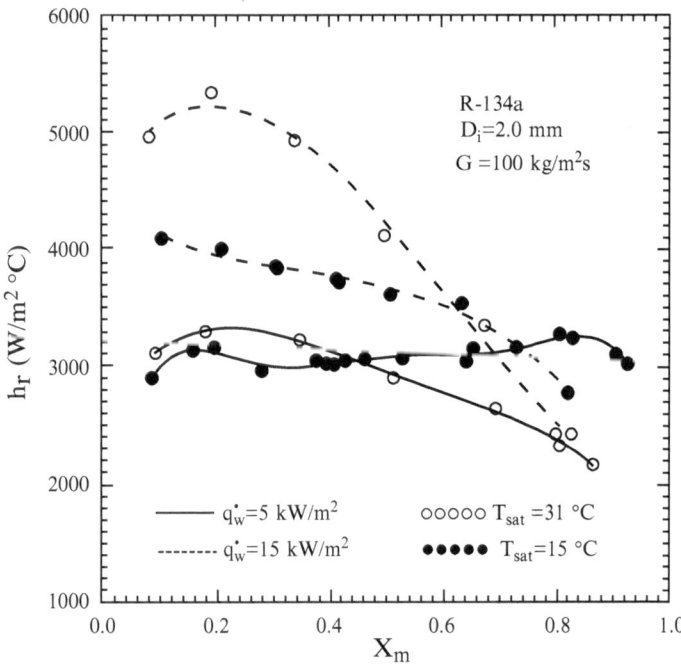

Fig. 2.2 Effects of the refrigerant saturated temperature on the evaporation heat transfer coefficient at two different wall heat fluxes [9]

Bulk fluid movement and the lateral movement of the bubbles in subcooled boiling are very important [17, 18]. Thin film dryout and receding contact angle during evaporation in a moving fluid should be considered as a mechanistic representation for CHF in flow boiling in microchannels.

2.2   Investigations of CHF

Many experimental investigations have been carried out [19–23]. There is a strong influence of flow regimes on CHF. Experiments on microchannels cover a wide range of geometry, physical dimensions, working fluids and operating parameters. There is a wide range of parameters [24, 25]. Very recent and important experimental investigations are given in Table 2.1 [26].

In case of microtubes; both single and multiple tubes have been used [27–31]. On the other hand, microchannels are studied in multiple. Parallel microchannel heat sinks are made of silicon, copper and stainless steel. Working fluids are water, refrigerants (R134a, R245fa, R236fa, R123, R32, R113 etc.), CO_2, nitrogen, helium, ethanol etc. CHF depends on type of fluid [32, 33]. However, CHF is a weak function of saturation.

CHF depends on the hydraulic diameter and length to diameter. For subcooled boiling, CHF increases with the decrease in channel diameter due to decrease in the departure diameter of the vapor bubbles, increase in bubble velocity relative to liquid and strong condensation at the tip of bubble [34, 35]. On the other hand for $x_{eq} > 0$, CHF decreases with decrease in d. CHF increases monotonically with the increase in mass flux and linearly with the increase in subcooling [36].

2.3   Prediction of CHF

The previous section discussed some important experimental investigations on flow boiling through microchannels. Some trends of the variation of CHF have already emerged. This has encouraged the researchers to try for the prediction of CHF as a function of geometric and operational parameters as well as fluid properties. Such predictions can be made through correlations or modeling. In this section, the correlations for the CHF in microchannels have been examined. Table 2.2 provides a compilation of correlations relevant for the CHF in microchannel [26]. Critical heat flux varies proportionally with the mass velocity and the enthalpy needed for vaporization [37–40].

Katto [37] predicted the correlation for uniformly heated vertical channels. Wu et al. [41] used the Katto [37] correlation for a large number of refrigerants, water and nitrogen, and obtained a reasonable prediction. The correlation by Katto and Ohno [42] has been used for predicting saturated CHF in a single channel.

Table 2.1 Some typical investigations CHF during flow boiling through micro channels [26]

Author	Geometry	No. of channels	Fluid	Test conditions	Remark
Mauro et al. (2010)	199 μm wide and 756 μm deep (**copper heat sink**)	29	R134a, R236fa, R245fa	Mass fluxes = 250–1,500 kg/m² s inlet subcooling from 25 to 5 K saturation temperatures = 20–50 °C	Effect of subcooling is not so much, mass flux increases CHF increases. Effect of saturation temp ~ effect of system pressure
Park and Thome (2010)	(1) 467 μm wide and 4,052 μm deep, (2) 199 μm wide and 756 μm deep; total length = 30 mm, heated length = 20 mm (**copper heat sink**)	(1) 20 (2) 29	R134a, R236fa, R245fa	Mass flux = 100–400 kg = m² s; inlet subcooling from 20 to 3 K	For low mass velocity all the refrigerants showing same CHF. Depending upon CHF and pressure drop, R134a is best coolant than R236fa and 245fa
Roday et al. (2009)	Microtube, inner diameter = 0.286–0.700 mm, total length = 121.96–138.65 mm, heated length = 21.66–90.84 mm (**stainless steel**)	Single	Water, R-123	Mass flux = 320–1,570 kg/m² s, exit pressure = 25.3–225 kPa, subcool temperature = 2–80 °C	CHF dependent on mass flux, heated length, channel diameter, exit pressure as well as liquid subcooling
Agostini et al. (2008)	Parallel channels (223 μm wide, 680 μm high and 20 mm long with 80 lm thick fins separating the channels) (**silicon**)	67	R236fa, R245fa	Mass flux = 276–992 kg/m² s, subcool temperature = 0.4–15 °C	CHF increases with mass flux. Effect of inlet subcooling and saturation temperature is negligible. Effect of exit pressure has not studied
Wojtan et al. (2006)	Microtube, inner diameter = 0.5, 0.8 mm, heated length = 20–70 mm (**stainless steel**)	Single	R-134a, R-245fa	Mass flux = 400–1,600 kg/m² s, subcool temperature = 4–12 °C	CHF dependent on mass flux, heated length, channel diameter but not on liquid subcooling, Qu-Mudawar correlation over predicts CHF for the working range. Pressure effect has not studied

Hetsroni et al. (2006)	Hydraulic diameter 220, 130, and 100 µm, (parallel triangular micro-channels) (silicon substrate) (silicon substrate)	13, 21, and 26	Water, ethanol	Mass flux 32–200 kg/m² s; heat flux 120–270 kW/m²; vapor quality 0.01–0.08	When liquid film thickness reaches minimum CHF occurs
Lee et al. (2005)	231 µm wide × 713 µm deep (copper block)	53	R-134a	Mass velocity = 127–654 kg/m² s	At exit quality = 0.55 CHF can be observed
Qu et al. (2003)	215 µm × 821 µm	21	Deionized water	Inlet temperature 30–60 °C; mass velocity 86–368 kg/m² s	With low mass flux no effect of inlet temperature
Stoddard et al. (2002)	Annuli; inner diameter = 6.45 mm. Gap = 0.724–1.001 mm. Heated length = 185 mm	Single	Water	Mass flux = 100–380 kg/m² s; exit pressure = 0344–1.034 MPa; wall heat flux: 0.231–1.068 MW/m	Comparison of vertical and horizontal identical channel has been done

There are fewer correlations for horizontal flow than those for vertical flow at low mass fluxes. Groeneveld [43] has suggested a simple method of prediction for CHF in horizontal channel by multiplying the corresponding value in vertical tubes with a constant. According to Yu et al. [44] this gives the correct trend of CHF in small tubes. Shah [45] has proposed a widely applicable correlation for CHF in uniformly heated vertical channel. Tong [46] derived a CHF correlation by applying boundary layer; one parameter was given as a function of quality. Nariai et al. [47] studied CHF of subcooled flow boiling in narrow tubes. Celata et al. [48] also modified Tong's correlation. Hall and Mudawar [49] derived a statistical five-parameter correlation.

Qu and Mudawar [38] compared the Katto-Ohno [42] correlation and found vapor backflow in the upstream plenum as a result of flow instabilities and CHF was independent of inlet subcooling. Wojtan et al. [40] developed a correlation from R-134a and R-245fa data. Ong and Thome [50] and Zhang et al. [51] used an extensive to develop the correlations.

Kosar and Peles [52] concluded that CHF decreased with exit quality. A similar trend was also observed by Qi et al. [53]. However, the Kosar and Peles [52] correlation has poor predictive ability. Qi et al. [53] proposed a correlation that has been verified for aqueous data only. Wu et al. [41] developed a complicated correlation.

An excellent survey for predictive correlations for CHF has been done by Revellin et al. [54]. For saturated boiling CHF increases with mass flux with hydraulic diameter, heated length, subcooling and saturation pressure.

2.4 Models

Our knowledge of the nature of CHF is still not complete [55, 56]. An important aspect of multi-microchannel block is the conjugate effect. A multiple microchannel system is often considered as a single heat sink. Such channels at relatively high mass flux have been studied [43, 57, 58]. Though, most of the microchannel applications are limited to not-very-high mass flux, care must be taken in their design and operation as the CHF behavior can deviate substantially from the usual trend. There could be a number of flow regimes. A unique phenomenon related to microchannel flow boiling is the rapidly expanding vapor bubbles. Also, there may be a reverse flow of the liquid in the direction of the inlet manifold.

2.5 Boiling and CHF studies

CHF studies can be grouped into studies in single and parallel microchannels. Jiang et al. [59] investigated phase-change in multiple microchannel heat sink systems. Yen et al. [60] used single circular tubes and observed that the exit quality at CHF was approximately 1.0.

Table 2.2 Correlations used for the prediction of CHF in microchannels [26]

Sl. no.	Correlation	Author(s)	Details	Error	Remarks
1	$Bl_{chf} = 0.10(\rho_v / \rho_l)^{0.133} (1/We_{lo})^{0.333} \dfrac{1}{1+0.03(L_h / d_h)}$	Katto (1978)	Circular, conv. channel	Percentage of data points within the 30 % error band: 40.7 %, MEA=46 % SD: 69.1 %	Originally developed for macrochannels. However, different authors have compared it against microchannel experimental data and it has fared well. Wu et al. (2011) have compiled the results from a microchannel database drawn from various sources
2	$q_{chf} = Gh_{lv}q_{Co}\left(1 + k\dfrac{\Delta h_{sub}}{h_{lv}}\right)$	Katto and Ohno (1984)		*Aqueous*: No. of data points: 2,427; percent of data points within the 30 %; error band: 69.1 %; MAE: 27.9 %; MRE: 3.2 % *Non-aqueous*: No. of data points: 569; percent of data points within the 30 %; error band: 69.1 %; MAE: 26.3 %; MRE: 0.2 %	

(continued)

Table 2.2 (continued)

Sl. no.	Correlation	Author(s)	Details	Error	Remarks
3	$q_{hor}^* = k_{hor} \cdot q_{tab}^*$. **From a look-up Table.** $k_{hor} = \begin{cases} 0.0, G \leq G_{min}; \\ (G - G_{min})/(G_{max} - G), G_{min} < G < G_{max}; \\ 1.0, G \geq_{max} . \end{cases}$ $G_{min} = \dfrac{\sqrt{g D \rho_g (\rho_l - \rho_g)}}{x} \left(\dfrac{1}{0.65 + 1.11 x_u^{0.6}} \right)^2 ,$ $G_{max} = \left[\dfrac{g D^{1.2} \rho_l (\rho_l - \rho_g)}{0.092(1 - x_g)^{1.8} \mu_l^{0.2}} \left\{ \dfrac{-0.3470 + 0.092 \ln(x_g)}{-0.0556 \ln^2 (x_u)} \right\} \right]^{-0.056} .$	Groeneveld (1986)	Correlations on flow through horizontal channels as this are scarce, especially at low mass fluxes. According to Yu et. al. (2002), this formula gives the correct trend of CHF in small tubes		
4	$q_{chf} / G h_{fv} = 0.124 \left(\dfrac{L}{d} \right)^{-0.89} \left(\dfrac{10^4}{Y_{shah}} \right)^n (1 - x_{in})$ $Y_{shah} = G^{1.8} d^{0.6} \left(\dfrac{C_p}{K_l \rho_l^{0.8} g^{0.4}} \right) \left(\dfrac{\mu_l}{\mu_v} \right)^{0.6}$ If $Y_{shah} \leq 10^4, n = 0.$ If $10^4 < Y_{shah} \leq 10^6, n = \left(\dfrac{d}{L} \right)^{0.54}$ If $Y_{shah} > 10^6, n = \left(\dfrac{0.12}{(1 - x_{in})^{0.5}} \right)$	Shah (1987)	Derived from a database containing 23 different fluids and diameter range 0.315–37.5 mm	*Aqueous:* No. of data points: 2,427 Percent of data points within the 30 % error band: 76.3 % MAE: 25.0 % MRE: 11.3 % *Non-aqueous:* No. of data points: 569 Percent of data points within the 30 % error band: 41.7 % MAE: 33.7 MRE: −14.4	

#	Formula	Reference	Conditions	Results	Notes
5	$\dfrac{C}{C_{Tong}} = 1 - \dfrac{52.3 + 80 x_{eq,o} - 50 x_{eq,o}^2}{60.5 + (10 p_0)^{1.4}}$	Inasaka and Nariai (1987)	Can verify CHF data within ±20 % accuracy for channels down to 2 mm diameter	Mean deviation: 30.5 %	
6	$Bo = \dfrac{C}{\sqrt{Re}} = \dfrac{q_c}{G h_{fg}}$	Celata et al. (1994)	Can verify CHF data within a rms error of 21.2 % for channels down to 0.3 mm diameter	Mean deviation: 30.1 %	
7	$Bo = \dfrac{c_1 We_D^{c_2} (\rho_f/\rho_g)^{c_3} \left[1 - c_4 (\rho_f/\rho_g)^{c_5} x_{eq,in}\right]}{1 + 4 c_1 c_4 We_D^{c_2} (\rho_f/\rho_g)^{c_3+c_5} (L/d_h)}$; $c_1 = 0.0722; c_2 = -0.312; c_3 = -0.644; c_4 = 0.900; c_5 = 0.724$	Hall and Mudawar (2000)		Mean deviation: 19.2 %	
8	$Bl_{chf} = 33.43 (\rho_v/\rho_l)^{1.11} (1/We_{lo})^{0.21} (L_h/d_h)^{-0.36}$	Qu and Mudawar (2004)	Rectangular channel, hydraulic dia = 0.38–2.54 mm	Percent of data points within the 30 % error band: 37.6 % MAE: 44.9 %; SD: 56.4 % **Aqueous:** No. of data points: 2,427 Percent of data points within the 30 % error band: 1.9 % MAE: 418.4 %; MRE: 330.1 % **Non-Aqueous** No. of data points: 569 % of data points within the 30 % error band: 4.2 % MAE: 1,883.9 %; MRE: 1,883.3 %	Average of results obtained by nine author compiled by Wu et. al. (2011) Based on average of results obtained by ten author compiled by Revellin et al. (2009) Based on average of results obtained by 12 author compiled by Revellin et. al. (2009)

(continued)

Table 2.2 (continued)

Sl. no.	Correlation	Author(s)	Details	Error	Remarks
9	$Bl_{chf} = 0.437 \left(\rho_v / \rho_l\right)^{0.073} \left(1/We_{lo}\right)^{0.24} \left(L_h / d_h\right)^{-0.72}$	Wojtan et al. (2006)	Circular channel. $Dh = 0.5–0.8$ mm; valid for R134a, R245fa etc.	Percent of data points within the 30 % error band: 66.7 % MAE: 46.1 %; SD: 105.1 % **Aqueous:** Percent of data points within the 30 % error band: 27.7 % MAE: 95.0 %; **MRE:** 64.1 % **Non-Aqueous:** No. of data points: 569 Percent of data points within the 30 % error band: 64.9 % MAE: 25.1 %; **MRE:** 15.8	Based on the average of results obtained by eight authors, compiled by Wu et al. (2011) Based on the average of results obtained by ten authors, compiled by Revellín et al. (2009) Based on the average of results obtained by 12 authors, compiled by Revellín et al. (2009)
10	$\dfrac{q_{chf}}{Gh_{lv}} = 0.0352 \left[We_D + 0.0119 \left(\dfrac{L}{d}\right)^{2.31} \left(\dfrac{\rho_v}{\rho_l}\right)^{0.361} \right]^{-0.311}$ $\times 2.05 \left[\left(\dfrac{\rho_v}{\rho_l}\right)^{0.17} - x_{in} \right]$	Zhang et al. (2006)		**Aqueous:** No. of data points: 2,427 Percent of data points within the 30 % error band: 82.8 % MAE: 18.2 %; **MRE:** 2.7 % **Non-Aqueous:** No. of data points: 569 Percentage of data points within the 30 % error band: 63.6 % MAE: 31.7 %; **MRE:** 1.6 %	

#	Equation	Conditions	Results	Notes
11	$$\dfrac{q_c}{G h_{lv}} = 0.437 \left(\dfrac{\rho_v}{\rho_l}\right)^{0.073} We^{-0.24} \left(\dfrac{L_h}{d}\right)^{-0.72}$$	Katto–Ohno—modified (by Wotzan et al. in 2006)	Percent of data points within the 30 % error band: 82.4 % MAE: 7.6 %	
12	$$Bl_{chf} = \left\{ \left[0.0934 \left(P_e / P_{Cr}\right)^2 - 1.3 \times 10^{-4} \right] x_e^{0.59} \right\}^{1/1.08}$$	Kosar and Peles (2007)	Rectang. channel, $Dh \sim 0.228$ mm, for R123	Percent of data points within the 30 % error band: 26.6 %; MAE: 114.1 % SD: 146.7 %
13	$$Bl_{chf} = (0.214 + 0.14 Co)(\rho_v / \rho_l)^{0.133} \\ (1/We_{lo})^{0.333} \dfrac{1}{1 + 0.03 (L_h / d_h)}$$	Qi et al. (2007)	Cir. channel hyd. Dia. $= 0.53–1.93$ mm	Percent of data points within the 30 % error band: 8.3 % MAE: 313.4 % SD: 312.4 % **_Aqueous:_** No. of data points: 2,427 Percent of data points within the 30 % error band: 21.8 % MAE: 169.6 %; MRE: 147.5 **_Non-Aqueous:_** No of data points: 569 Percent of data points within the 30 % error band: 13.2 % MAE: 206 % MRE: 203.9 %	Based on the average of results obtained by eight authors, compiled by Wu et al. (2011) Based on average of results of ten authors, compiled by Revellin et al. (2009)

(continued)

Table 2.2 (continued)

Sl. no.	Correlation	Author(s)	Details	Error	Remarks
14	$q_{chf} = q_{co}\left(1 + K'\,\dfrac{\Delta h_{sub}}{h_{lv}}\right)$ If $\dfrac{1}{We_{lo}} < 3\times10^{-6}$, then $K' = 1.8\left(\dfrac{L}{130d}\right)^{\frac{5\rho_x}{\rho_l}}$; else $K' = 0.075\left(\dfrac{L}{130d}\right)^{\frac{5\rho_x}{\rho_l}}We_{lo}^{0.25}$.	Qi et al. (2007)			
15	$\dfrac{q_{chf}}{Gh_{lv}} = 0.12\left(\dfrac{\rho_v}{\rho_l}\right)^{0.062}We^{-0.141}\left(\dfrac{L_h}{d}\right)^{-0.7}\left(\dfrac{\mu_l}{\mu_v}\right)^{0.183}\left(\dfrac{d}{d_h}\right)^{0.11}$	Ong and Thome (2011)		No of points: 21 Percent of data points within the 30 % error band: 94.4 % MAE: 13.6 %	
16	$Bl_{chf} = q_{chf}/(Gh_{lv}) = 0.62\left(L_h/d_h\right)^{-1.19}x_e^{0.817}$	Wu et al. (2011)	Valid for five fluids incl. water	This is based on 629 (133 aqueous, 469 non-aqueous) data points, covering 5 halogenated refrigerants, nitrogen and water	It predicts almost 97 % of the non-aqueous data (except R12) and 94 % of water data within the ±30 % error data band. For non-aqueous data, the MAE and SD are 11.1 % and 13.9 % respectively. For water, they are 17.2 % and 18.4 % respectively

17	$Bl_{chf} = 0.6\left(L_{h}/d_{h}\right)^{-1.2} x_{e}^{0.82}$	Wu and Li (2011)	Percent of data points within the 30 % error band: 96.5 % MAE: 10.2 % SD: 13.7 % MRE: 2.4 %	Region-I: (726 data points) $L_{h}/d_{h} \leq 150$, $BoRe_{l}^{0.5} \leq 200$
18	$Bl_{chf} = 1.16\times10^{-3}\left(We_{m} Ca_{l}^{0.8}\right)^{-0.16}$	Wu and Li (2011)	Percent of data points within the 30 % error band: 82.0 % MAE: 16.1 % SD: 20.5 % MRE: –5.9 %	Region-II: (133 points) $L_{h}/d_{h} > 150$, $BoRe_{l}^{0.5} \leq 200$.
	Key: MAE: mean absolute error, MRE: mean relative error, SD: standard deviation, rms: root mean square			

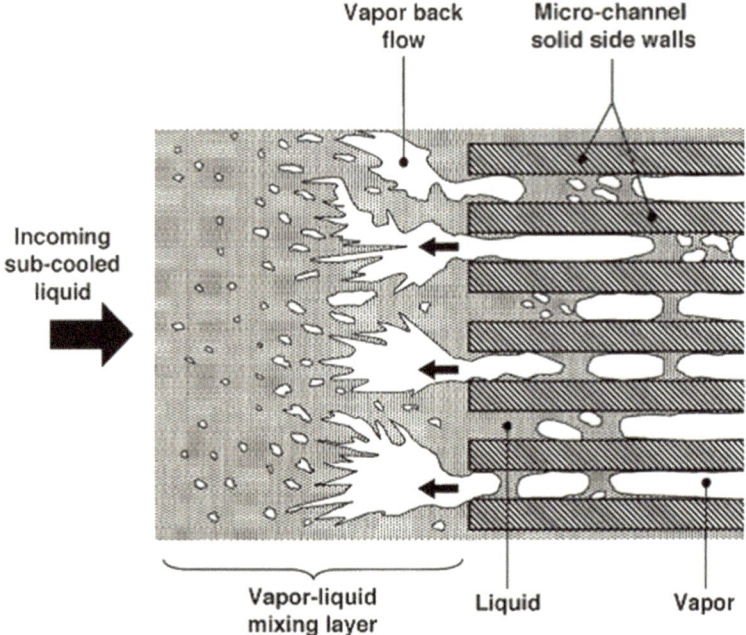

Fig. 2.3 Vapor back flow in the experiments of Qu and Mudawar [38]

Bowers and Mudawar [61] used R-113 in circular mini-channel and 510 μm diameter copper micro-channel heat sinks. They concluded that the CHF was not a function of inlet subcooling. CHF increased with mass velocity.

Qu and Mudawar [38] measured the CHF for a water-cooled microchannel heat sink that contained 21 parallel microchannels. They observed back flow, Fig. 2.3 [38]. The CHF condition advanced with an increase in mass velocity and it was independent of inlet temperature. They proposed a new correlation for CHF.

Three major flow instabilities affect CHF in microchannel heat exchangers—the upstream compressible volume instability, the excursive instability, and the parallel channel instability. The upstream compressible volume instability causes severe pressure drop oscillation leading to a premature CHF [62]. Upstream throttling in [38] took care of the compressible volume instability, but it was subject to excursive instability. This can be avoided by increasing the available pressure drop. In the case of parallel microchannels, improved fabrication techniques can be used to employ flow restrictions at the inlet of each channel to accomplish the pressure drop needed [63]. This technique was used by Kosar et al. [64]. The value of CHF increased with an increase in restrictor length. Using the same device as in [64], Koşar and Peles [65] studied the CHF condition of R-123 at exit pressures ranging from 227 to 520 kPa. CHF data were obtained over wide heat flux and mass flux range. Dryout was the leading CHF mechanism. CHF increased fairly linearly with mass flux.

Kuan and Kandlikar [66] studied the effect of flow boiling stability on CHF with R-123 in six parallel microchannels. They studied the effect of using pressure drop elements to restrict the flow and reduce vapor backflow. There is decrease in the CHF value with the use of restrictors.

Conjugate heat transfer effects may become important in microchannels since the channel wall thickness becomes comparable to the microchannel size. For microchannels fabricated on large blocks, it is possible that there is significant axial and longitudinal conduction through the substrate. This results in redistribution of heat flux away from those locations where the CHF condition typically initiates i.e., at the exit and eventually leads to the CHF condition occurring at the hottest surface. Higher values of the apparent CHF could be obtained due to these conjugate effects. Conjugate heat transfer with single-phase flow in microchannels has been studied [67] numerically, but such studies for two-phase flow and the effects on CHF have not been reported. There are very few CHF investigations in small circular tubes [34, 68–70]. Bergles and Rohsenow [70] studied CHF with de-ionized. At high subcooling CHF decreased monotonically with increases in quality and then increased in the bulk boiling region following a minimum.

Roach et al. [71] studied the CHF associated with flow boiling of subcooled water in circular tubes. Oh and Englert [72] conducted sub-atmospheric CHF experiments with water in a single rectangular aluminum channel heated on one side with electric strip heaters. CHF experiments were performed by Lazarek and Black [73] with R-113 in a stainless steel. CHF occurred because of the dryout of the liquid and always at the exit of the heated test section length. Yu et al. [74] carried out CHF experiments with water in a stainless steel channel. The relative size of the channel compared to the wall thickness played a role in the CHF condition.

Lezzi et al. [75] reported experimental results on CHF in forced convection boiling of water in a horizontal tube and the critical heat flux was reached due to dryout. They claimed that no oscillations affected the CHF condition. Wojtan et al. [40] investigated saturated critical heat flux in single uniformly heated microchannels with R-134a and R-245fa. They presented a new correlation to predict CHF in circular uniformly heated microchannel. Harirchian and Garimella [76, 77] reported five major flow regimes; bubbly, slug, churn, wispy annular, annular and a post dryout regime of inverted annular flow. The observed flow regimes were not much different from those observed in large size conduits.

A scale analysis based theoretical model for CHF was proposed by Kandlikar [78]. He considered evaporation momentum, surface tension, inertia and viscous forces and the constants were extracted from the available experimental data. Dryout during flow boiling has been associated with the elongated bubbles, [79] or annular flow regime. Kandilkar [80] observed the local dry-patch in elongated bubbles, micro-layer evaporation and the meniscus of an expanding bubble.

Revellin and Thome [81] suggested a mechanistic mode of CHF for flow boiling through heated microchannels. Revellin and Thome [24] extended the film dryout model. They argued that even though water had higher CHF compared to all other fluids, it is not good for electronic component cooling due to its very low saturation pressure at 30–40 °C. Revellin et al. [82] have done optimization analysis using CHF model on a constructal based tree shaped microchannel network in a disc shaped heat sink.

Kosar [83] constructed a simple model of CHF for saturated flow boiling. Kuan and Kandlikar [84] proposed another mechanistic model of critical heat flux based on force balance at the interface of a vapor plug in a microchannel. Yen et al. [85] studied convective flow boiling in a circular Pyrex glass microtube and a square Pyrex glass microchannel. Higher heat transfer coefficient was observed in the square microchannel as compared to the circular cross sectional microtube because of square corners acting as active nucleation sites. The reader is advised to refer [86–113] for more information.

2.6   Some Effects on CHF

2.6.1   Effect of Mass flux on CHF

The CHF increases with increase in mass flux Fig. 2.4 [38] and exit pressure [40, 63, 65, 114]. CHF also increases with smaller hydraulic diameter and conjugate effects. Slope of CHF vs. mass flux curve also depends on pressure.

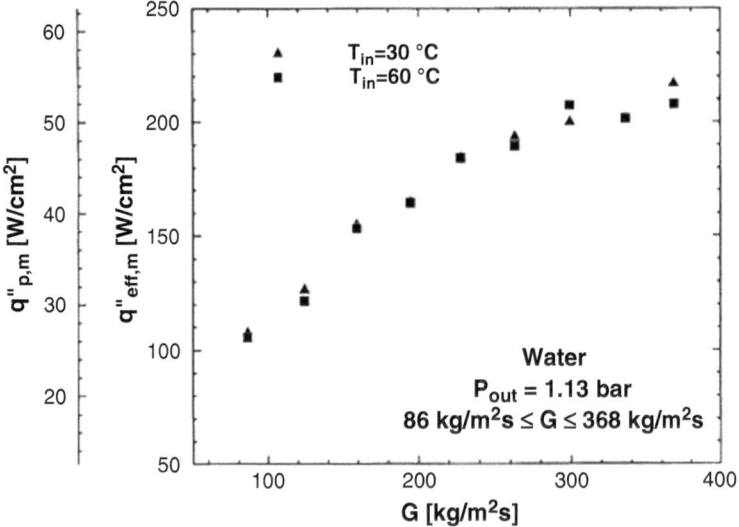

Fig. 2.4   Effect of mass flux on CHF [38]

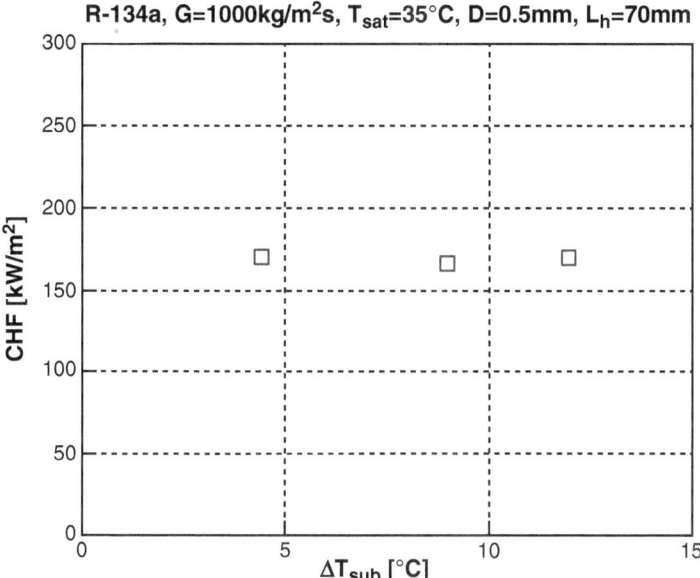

Fig. 2.5 Inlet subcooling effect on CHF [40]

2.6.2 Effect of Inlet Subcooling

At high mass flux, for high inlet subcooling, CHF decreases with decrease in inlet subcooling. But for low subcooling, the CHF increases with decrease in subcooling. The exit quality at CHF is close to zero. Figure 2.5 shows effect of inlet subcooling on CHF [40]. Parallel channel instability occurs [38] with the approach of CHF since then vapor mixes with subcooled inlet fluid in the plenum and the inlet subcooling loses its influence [40].

2.6.3 Effect of Exit Quality on CHF

The CHF increases with quality [115], Fig. 2.6 and CHF is high in the region close to saturation when compared to the high subcooled region due to change in void fraction and the flow velocity. References [115–117] may be read for more information.

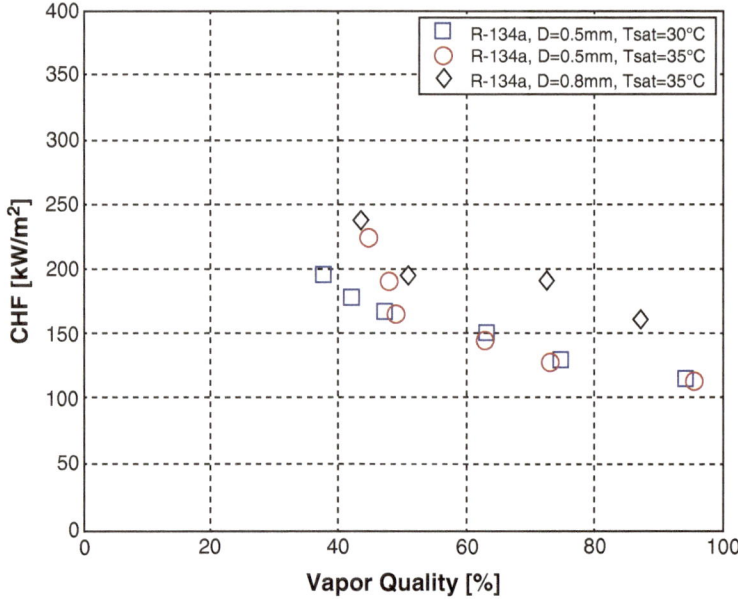

Fig. 2.6 Variation of CHF with exit quality [40]

2.6.4 Effect of Tube Diameter on CHF

CHF increases substantially with a reduction in tube diameter, [40, 63, 70, 114], Fig. 2.7 [40].

2.6.5 Effect of Heated Length on CHF

CHF decreases with an increase in heated length [40, 118].

Note: References [119–139] may be read for more information.

2.7 Some Important Results and Observations

As shown in Fig. 2.8, critical site number preventing the liquid supply to the micro-layer under the bubble is to be determined before the CHF is evaluated [15]. Figure 2.9 shows effect of surface wettability on CHF [15]. The active nucleation density, the bubble departure diameter and their product are affected by contact angle which in turn affects CHF.

Figure 2.10 compares the predicted and measured fraction of dry area near CHF [15]. The estimated fraction of dry out area at CHF is not constant. It depends

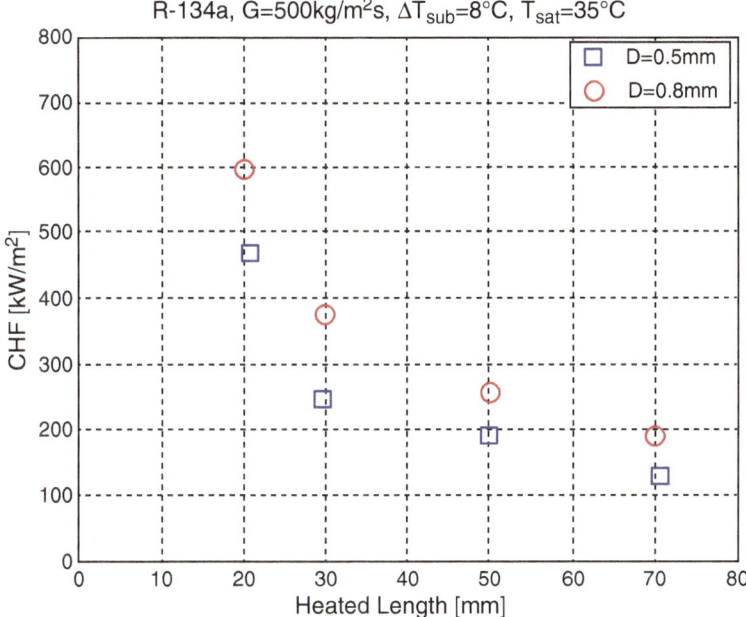

Fig. 2.7 Variation of CHF with heated length [40]

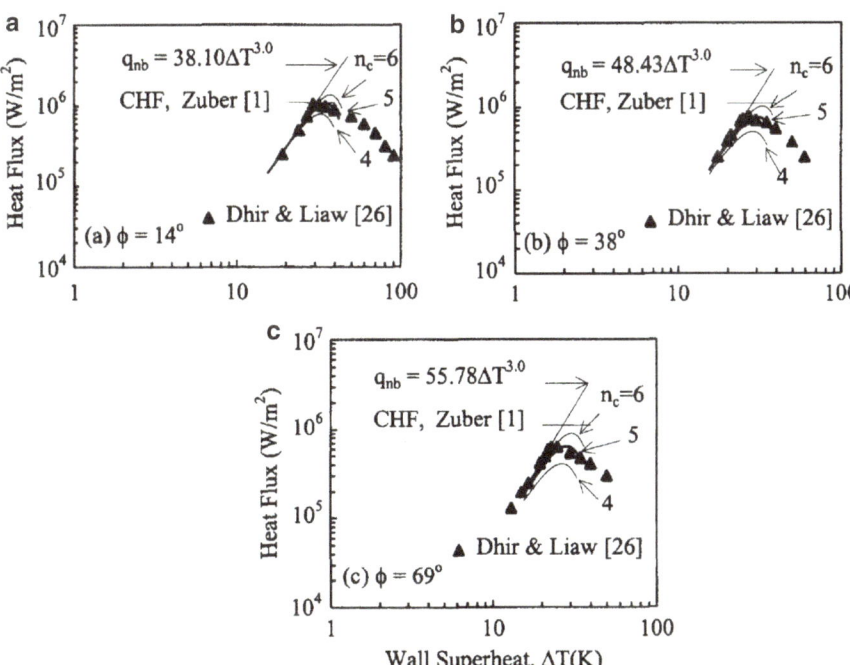

Fig. 2.8 Comparison of predictions by assuming critical site number and the experimental data for several contact angles [15]

Fig. 2.9 Effect of surface
wettability on CHF [15]

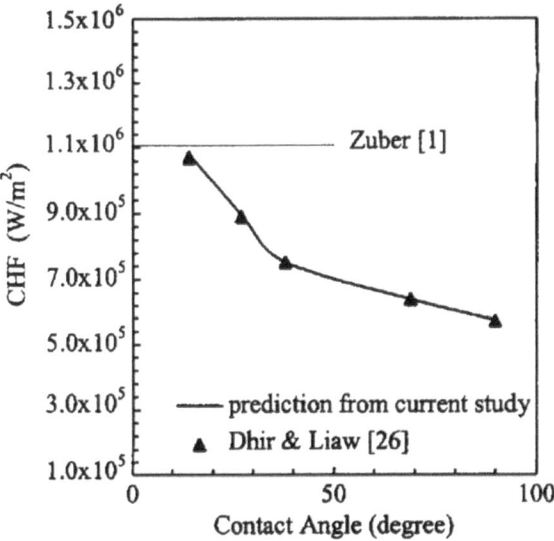

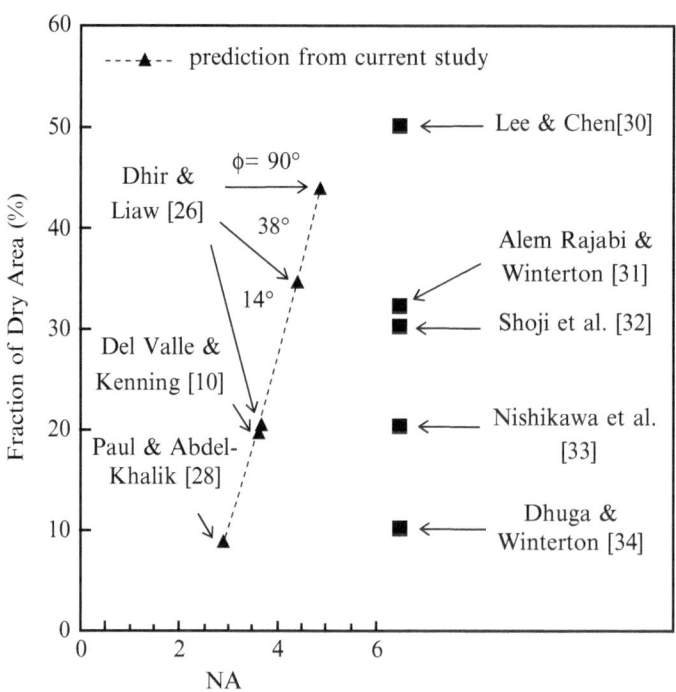

Fig. 2.10 Comparison of predicted and measured fraction of dry area near CHF [15]

on boiling condition. The fraction of dry area is high for the high wall void fraction at CHF. Figure 2.11 shows comparison between experimental and calculated CHF at inlet conditions with contact angle 50° [16]. The average predicted to measured CHF ratios (CHFR) decrease with the increase in contact angle. In this study [16], a dry spot model of CHF has been developed for both pool boiling and subcooled forced convection boiling. Figure 2.12 shows the CHF behavior with quality based

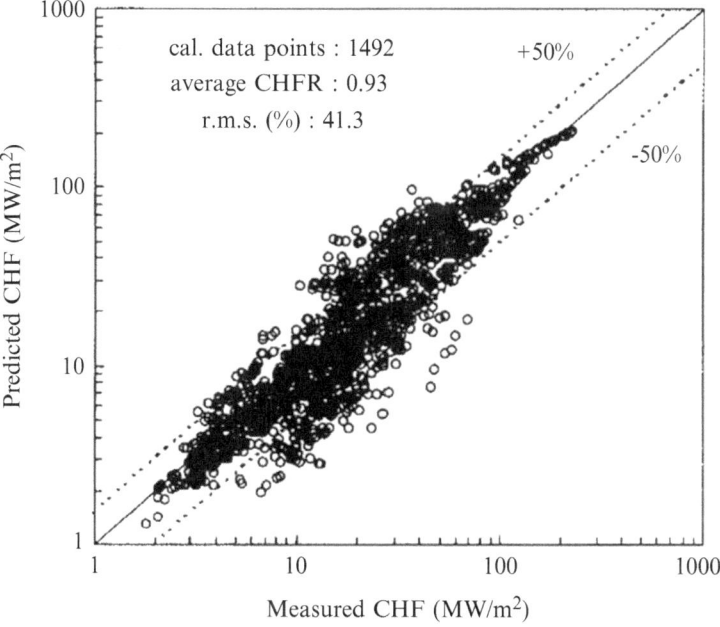

Fig. 2.11 Experimental vs. calculated CHF at contact angle 50° [16]

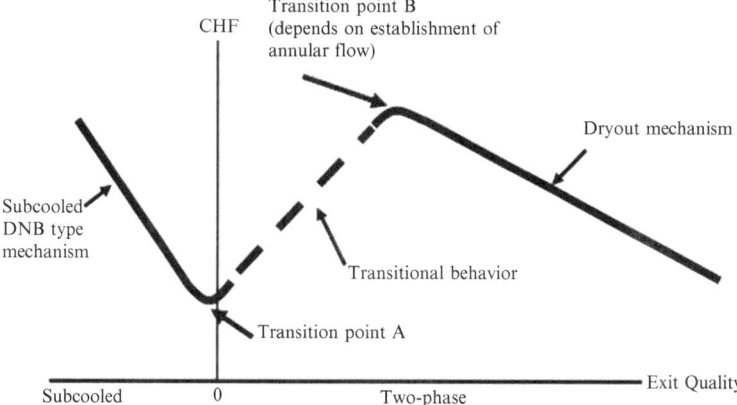

Fig. 2.12 Speculation on the trends of CHF with exit quality [29]

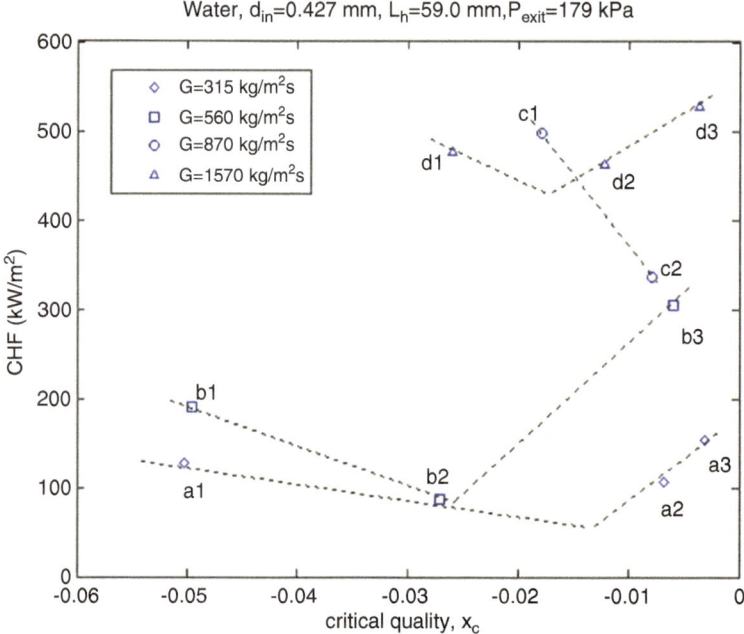

Fig. 2.13   CHF variations with quality in the subcooled region [30]

on the flow regime map. There are two transition points and these transitions depend on the condition of operating parameters. The mechanism changes from DNB type behavior in the subcooled region to dryout behavior at high qualities [29]. For critical qualities below point of net vapor generation (PNVG), CHF decreases with quality and for above PNVG, CHF increases with quality as shown in Fig. 2.13 [30].

It is important to identify CHF conditions based on the criteria for such description and the repeatability of the measurements must be ensured. The slope of the heat flux vs. temperature curve is high at the beginning of two-phase region and the gradient decreases as the boiling crisis approaches. Figure 2.14 [32] shows this and, in such a way, the integrity of the test setup is ensured.

Figure 2.15 shows the curves for CHFs with mass velocity [33]. While CHF increases with mass velocity, its rate of rise is less at high mass velocities. CHF increases moderately with increasing inlet subcooling.

CHF virtually does not depend on dissolved gas concentration from near zero to the saturation level as shown in Fig. 2.16, [34]. The CHF correlation for a horizontal channel may be obtained by multiplying a constant as shown in Fig. 2.17 [36].

With heat flux approaching CHF, the intense parallel channel instability causes vapor backflow and mixing of vapor in the incoming subcooled liquid, the slope of the boiling curve increases indicating flow boiling near the outlet. However, when the heat flux approaches CHF, the slope again decreases and the heat transfer becomes less effective, Fig. 2.18 [38].

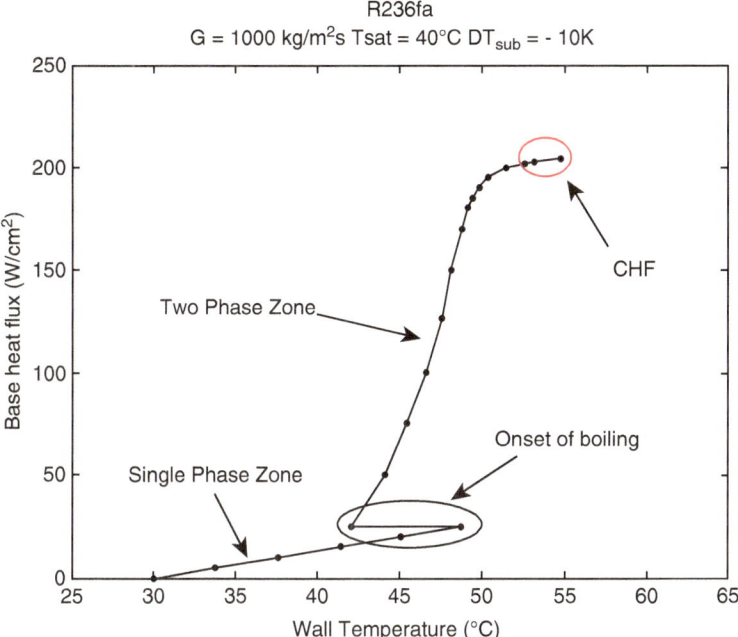

Fig. 2.14   Boiling curves [32]

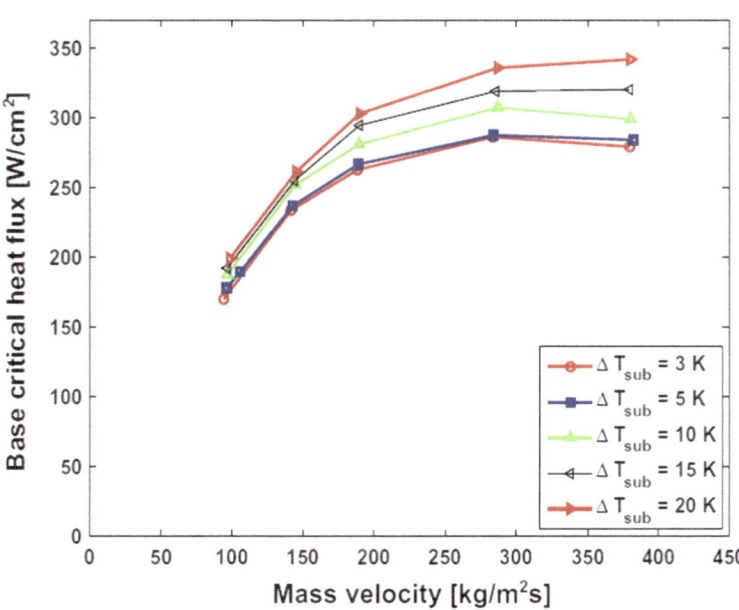

Fig. 2.15   CHF with mass velocity [33]

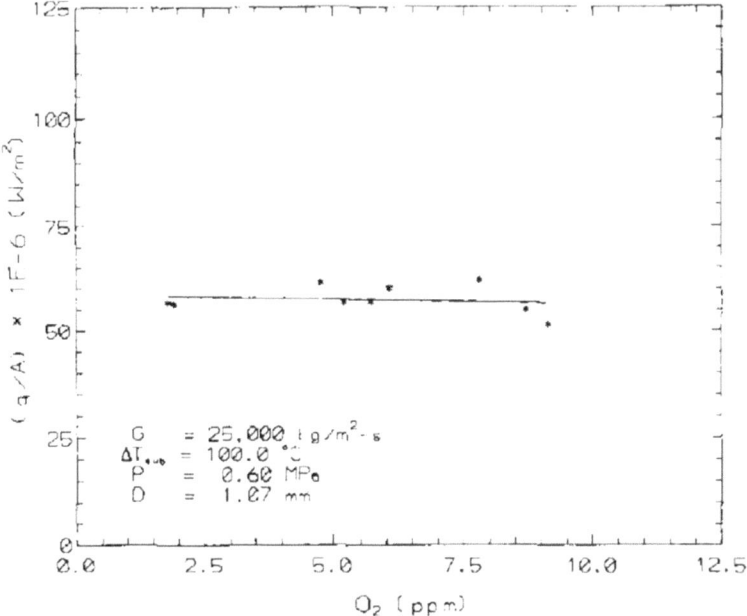

Fig. 2.16 CHF vs. dissolved oxygen concentration [34]

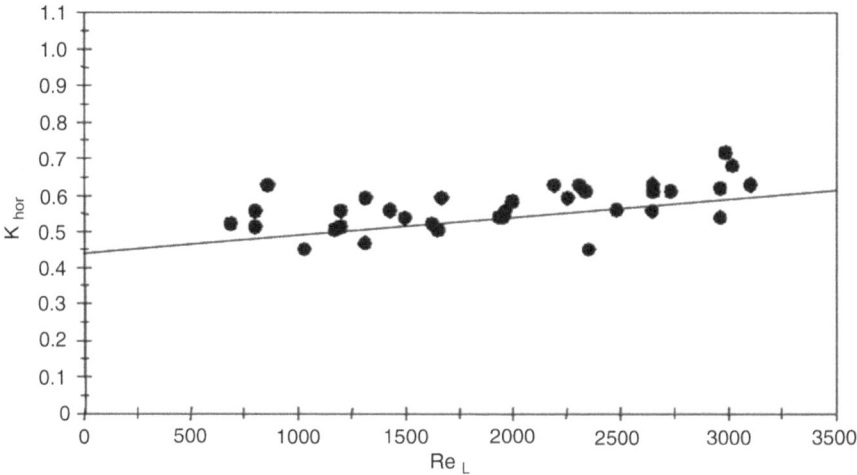

Fig. 2.17 The correction factor [36]

Flashing evaporation causes non-linear variation of the mass quality along the microtube as shown in Fig. 2.19 [39]; mass quality increases rapidly near the outlet. The evolution of the heat flux is shown in Fig. 2.20 [40]. Initially, the heat flux increases linearly with a small increase of the wall temperature; reaches the maximum when lot of vapor forms and the liquid becomes unable to wet the surface

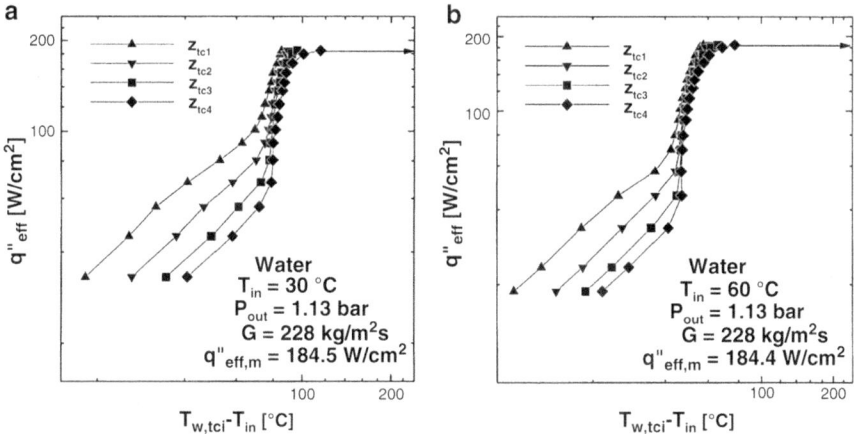

Fig. 2.18 Boiling curves [38]

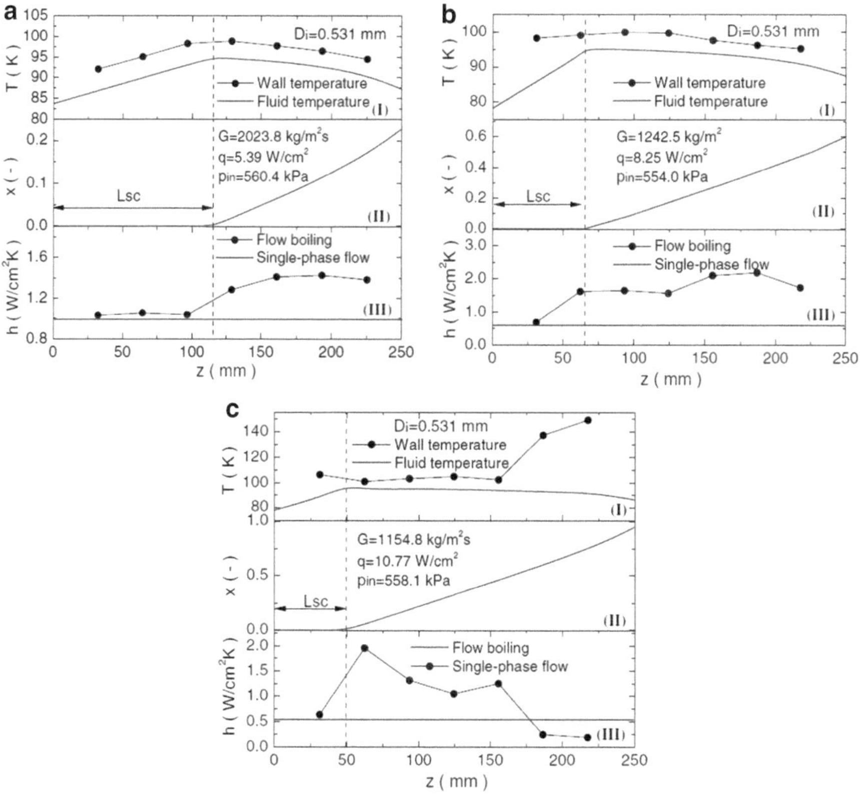

Fig. 2.19 Local flow boiling heat transfer characteristics at (**a**) low, (**b**) medium and (**c**) high heat flux [39]

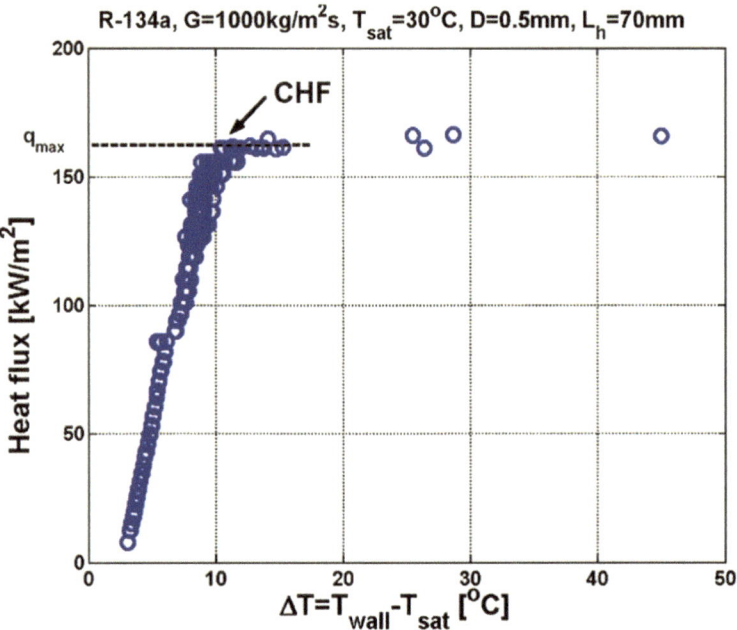

Fig. 2.20 Determination of CHF by experiment [40]

continuously. Consequently, heat transfer coefficient falls and the temperature rises and further heating must be stopped.

The variation of boiling number at CHF with different parameters is shown in Fig. 2.21 [41]. This is obtained from the data of different investigators. Figure 2.22 shows that the deviation of the predicted data for CHF from the correlations developed is often ±30 to ±50 % from the experimental data and the availability of

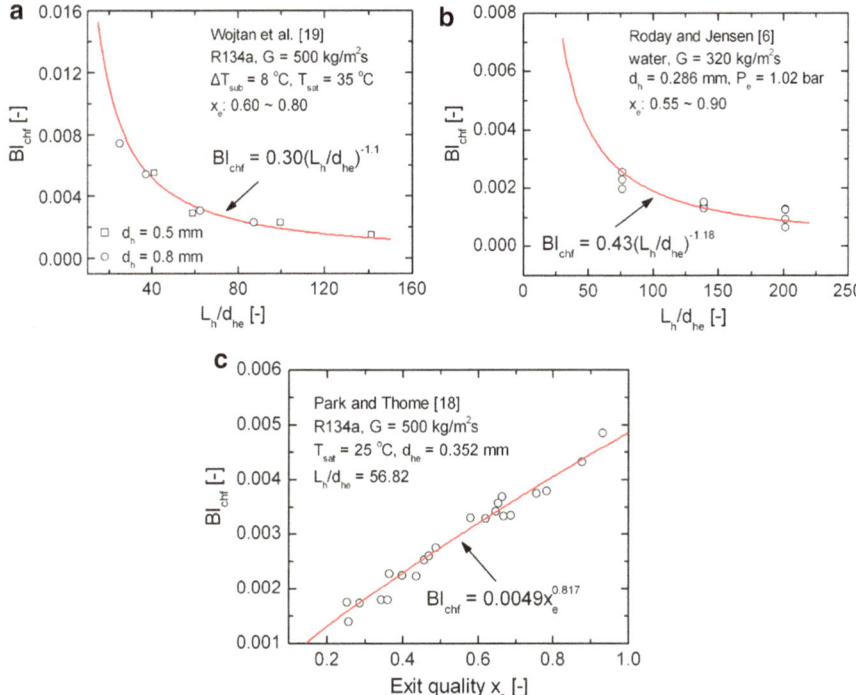

Fig. 2.21 Variation of boiling number at CHF with various parameters as obtained from the experimental data of different investigators [41]. (**a**) Wojtan et al.'s R134a data [140]. (**b**) Roday and Jensen's water data [141]. (**c**) Park and Thome's R134a data [109]

widely applicable correlation remains as a remote possibility [49]. Figures 2.23 and 2.24 [50] show the dependency of various parameters on pressures. Figures 2.25 and 2.26 [93] show confinement effects on CHF in buoyancy driven microchannels.

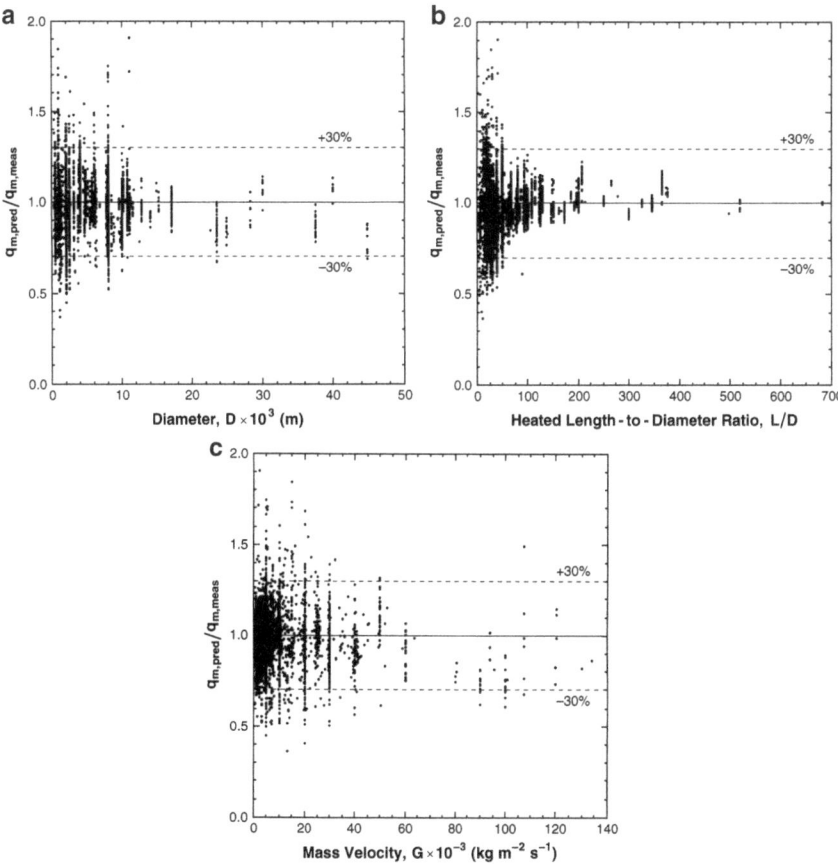

Fig. 2.22 Comparison of predicted CHF from the correlation with that from the experiment [49]

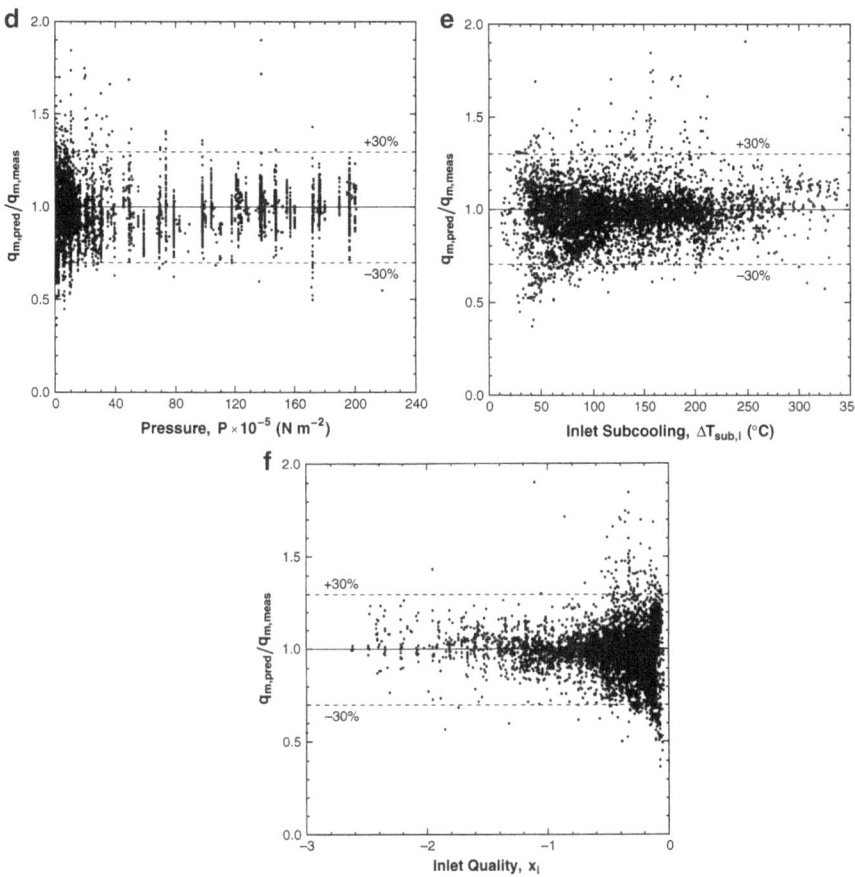

Fig. 2.22 (continued)

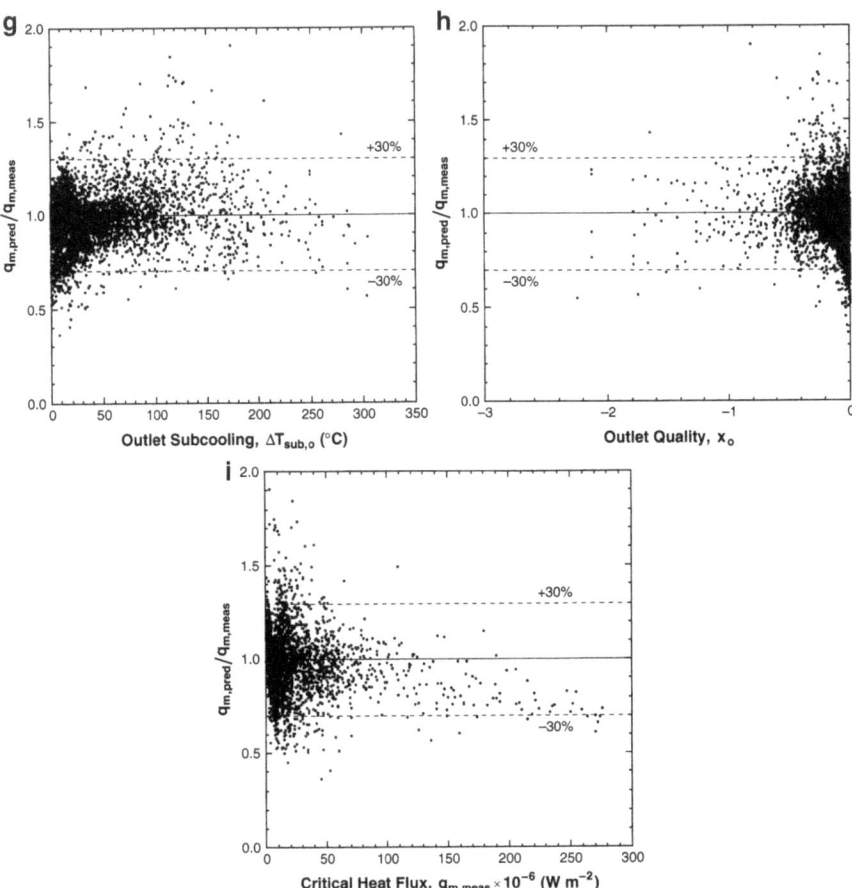

Fig. 2.22 (continued)

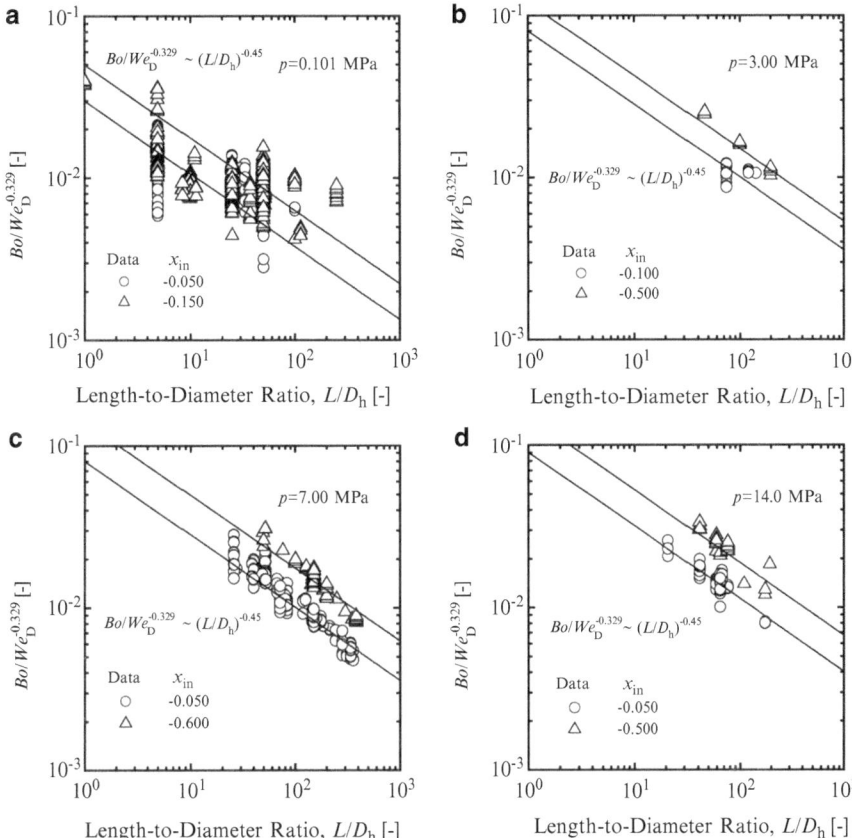

Fig. 2.23 Parameter dependencies with pressure [50]

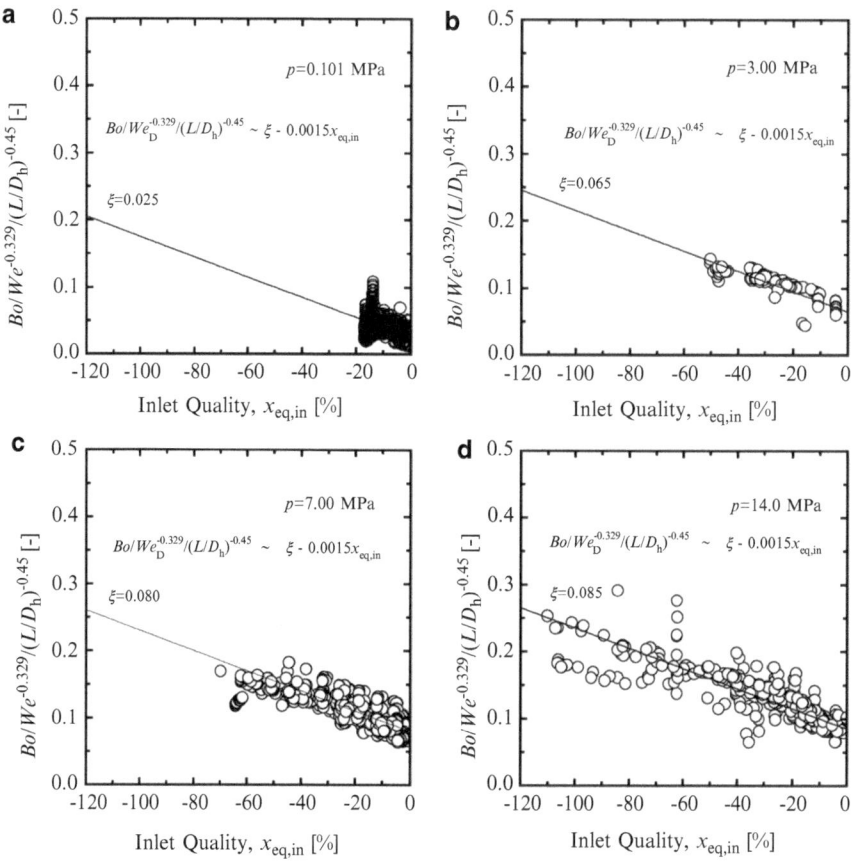

Fig. 2.24 Parameter dependencies with pressure [50]

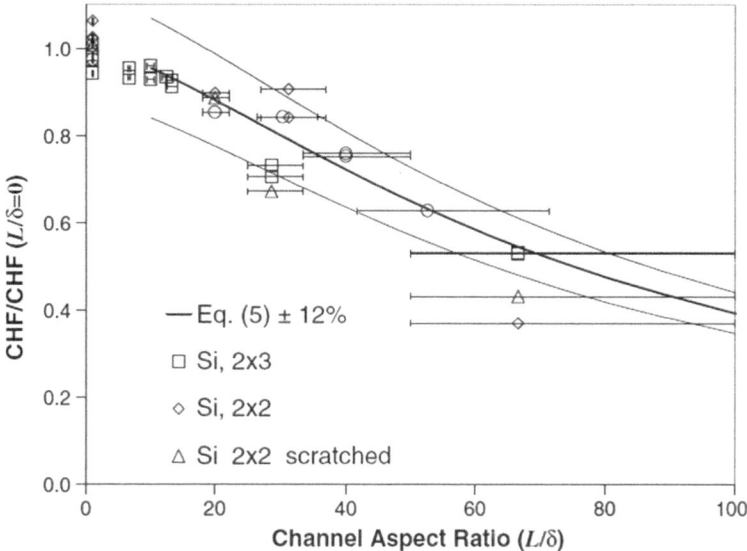

Fig. 2.25 CHF depends on channel aspect ratio for asymmetric heating [93]

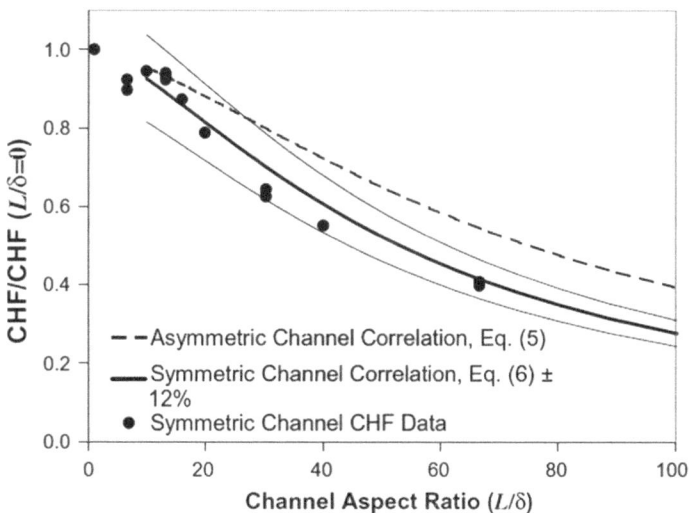

Fig. 2.26 CHF depends on channel aspect ratio for symmetric channel [93]

References

1. Collier JG, Thome JR (1994) Convective boiling and condensation, 3rd edn. Oxford Science Publications, New York, 1–33, 131–182, 183–213, 325–374
2. Hewitt GF (1998) Handbook of heat transfer, boiling, 3rd edn. McGraw-Hill, New York
3. Nukiyama S (1966) The maximum and minimum values of the heat Q transmitted from metal to boiling water under atmospheric pressure. Int J Heat Mass Transf 9:1419–1433
4. Katto Y (1994) Critical heat flux. Int J Multiph Flow 20(1):53–90
5. Tong LS, Tang YS (1997) Boiling heat transfer and two-phase flow, 2nd edn. Taylor & Francis, Bristol
6. Chang SH, Baek WP (2003) Understanding predicting and enhancing critical heat flux. In: The 10th international topical meeting on nuclear reactor thermo-hydraulics (NURETH-10), Seoul
7. Zuber N (1959) Hydrodynamic aspects of boiling heat transfer. PhD thesis, Research Laboratory, Los Angeles and Ramo-Wooldridge Corporation, University of California, Los Angeles
8. Lienhard JH, Dhir VK (1973) Extended hydrodynamic theory of the peak and minimum pool boiling heat fluxes. NASA CR-2270, contract no. NGL 18-001-035
9. Yan Y, Lin T (1998) Evaporation heat transfer and pressure drop of refrigerant R-134a in a small pipe. Int J Heat Mass Transf 41:4183–4194
10. Haramura Y, Katto Y (1983) A new hydrodynamic model of critical heat flux applicable to both pool and forced convection boiling on submerged bodies in saturated liquids. Int J Heat Mass Transf 26:389–399
11. Dhir VK, Liaw SP (1989) Framework for a unified model for nucleate and transition pool boiling. J Heat Transf 111(3):739–746
12. Liaw SP, Dhir VK (1989) Void fraction measurements during saturated pool boiling of water on partially wetted vertical surfaces. Trans ASME J Heat Transf 111(3):731–738
13. Bergles AE (1992) What is the real mechanism of CHF in pool boiling. In: Dhir VK, Bergles AE (eds) Pool and external flow boiling. ASME, New York, pp 165–170
14. Kandlikar SG (2001) A theoretical model to predict pool boiling CHF incorporating effects of contact angle and orientation. J Heat Transf 123(6):1071–1079
15. Ha SJ, No HC (1998) A dry-spot model of critical heat flux in pool and forced convention boiling. Int J Heat Mass Transf 41(2):303–311
16. Ha SJ, No HC (2000) A dry-spot model of critical heat flux applicable to both pool boiling and sub-cooled forced convention boiling. Int J Heat Mass Transf 43:241–250
17. Kandlikar SG (2001) Critical heat flux in sub-cooled flow boiling—an assessment of current understandings and future directions for research. Multiph Sci Technol 13(3):207–232
18. Celata GP, Mariani A (1999) CHF and post-CHF (post-dry-out) heat transfer, Chapter 17. In: Kandlikar SG, Shoji M, Dhir VK (eds) Handbook of phase change, boiling and condensation. Taylor and Francis, New York, pp 443–493
19. Bergles AE, Kandlikar SG (2005) On the nature of critical heat flux in micro-channels. J Heat Transf 127:101–107
20. Kim YH, Kim SJ, Noh SW, Suh KY et al (2003) Critical heat flux in narrow gap in two-dimensional slices under uniform heating condition. In: Transactions of the 17th international conference on structural mechanics in reactor technology (SMIRT 17), Prague, Czech Republic
21. Bar-Cohen A, Geisler K, Rahim E (2008) Pool and flow boiling in narrow gaps-application to 3D chip stacks. In: Proceedings of fifth European thermal-sciences conference
22. Aoki S, Inoue A, Aritomi M, Sakamoto Y (1982) Experimental study within on the boiling phenomena a narrow gap. Int J Heat Mass Transf 25(7):985–990
23. Kim JJ, Kim YH, Kim SJ et al (2004) Boiling visualization and critical heat flux phenomena in narrow rectangular gap. In: Fourth Japan-Korea symposium on nuclear thermal hydraulics and safety

24. Revellin R, Thome JR (2009) Critical heat flux during boiling in micro-channels: a parametric study. Heat Transf Eng 30(7):556–563
25. Ghiaasiaan SM, Abdul-Khalik SI (2001) Two phase flow in micro-channels. Adv Heat Transf 34:145–254
26. Das PK, Chakraborty S, Bhaduri S (2012) Critical heat flux during flow boiling in mini and microchannel—a state of the art review. Front Heat Mass Transf 3:013008
27. Roday AP, Jensen MK (2007) Experimental investigation of the CHF condition during flow boiling of water in micro-tubes. In: ASME-JSME thermal engineering summer heat transfer conference, Vancouver, Canada
28. Roday AP, Tasciuc TB, Jensen MK (2008) The critical heat flux condition with water in a uniformly heated micro-tube. J Heat Transf 130:1–9
29. Roday AP, Jensen MK (2009) Study of critical heat flux condition with water and R-123 during flow boiling in micro-tubes. Part I: experimental results and discussion of parametric effects. Int J Heat Mass Transf 52:3235–3249
30. Roday AP, Jensen MK (2009) Study of the critical heat flux condition with water and R-123 during flow boiling in micro-tubes. Part II: comparison of data with correlations and establishment of a new sub-cooled CHF correlation. Int J Heat Mass Transf 52(13–14): 3250–3256
31. Bower MB, Mudawar I (1994) High flux boiling in low flow rate, low pressure drop mini-channel and micro-channel heat sinks. Int J Heat Mass Transf 37(2):321–332
32. Mauro AW, Thome JR, Toto D, Vanoli GP (2010) Saturated critical heat flux in a multi-micro-channel heat sink fed by a split flow system. Exp Therm Fluid Sci 34:81–92
33. Park JE, Thome JR (2010) Critical heat flux in multi-micro-channel copper elements with low pressure refrigerants. Int J Heat Mass Transf 53:110–122
34. Vandervort CL, Bergles AE, Jensen MK (1994) An experimental study of critical heat flux in very high heat flux sub-cooled boiling. Int J Heat Mass Transf 37(Suppl 1):161–173
35. Bergles AE (1962) Sub-cooled burnout in tubes of small diameter. ASME paper 63-WA-182
36. Stoddard RM, Blasick AM, Ghiaasiaan SM, Abdel-Khalik SI, Jeter SM, Dowling MF (2002) Onset of flow instability and critical heat flux in thin horizontal annuli. Exp Therm Fluid Sci 26:1–14
37. Katto Y (1978) A generalized correlation of critical heat flux for the forced convection boiling in vertical uniformly heated round tubes. Int J Heat Mass Transf 21:1527–1542
38. Qu W, Mudawar I (2004) Measurement and correlation of critical heat flux in two-phase micro-channel heat sinks. Int J Heat Mass Transf 47:2045–2059
39. Qi SL, Zhang P, Wang RJ et al (2007) Flow boiling of liquid nitrogen in micro-tubes: part II: heat transfer characteristics and critical heat flux. Int J Heat Mass Transf 50:5017–5030
40. Wojtan L, Revellin R, Thome JR (2006) Investigation of saturated critical heat flux in a single, uniformly heated micro-channel. Exp Therm Fluid Sci 30:765–774
41. Wu Z, Li W, Ye S (2011) Correlations for saturated critical heat flux in micro-channels. Int J Heat Mass Transf 54:379–389
42. Katto Y, Ohno H (1984) An improved version of the generalized correlation of critical heat flux for the forced convective boiling in uniformly heated vertical tubes. Int J Heat Mass Transf 27:1641–1648
43. Groeneveld DC (1986) The onset of dry sheath condition—a new definition of dry-out. Nucl Eng Des 92:135–140
44. Yu W, France DM, Wambsganss MW, Hull JR (2002) Two-phase pressure drop boiling heat transfer and critical heat flux to water in a small-diameter horizontal tube. Int J Multiph Flow 28:927–941
45. Shah MM (1987) Improved general correlation for critical heat flux during up flow in uniformly heated vertical tubes. Int J Heat Fluid Flow 8:326–335
46. Tong LS (1968) Boundary-layer analysis of the flow boiling crisis. Int J Heat Mass Transf 11:1208–1211

47. Nariai H, Inasaka F, Shimuara T (1987) Critical heat flux of sub-cooled flow boiling in narrow tubes. In: ASME/JSME thermal engineering joint conference (1987), vol 5, pp 455–462
48. Celeta GP, Cumo M, Mariani A (1993) Burnout in highly sub-cooled water flow boiling in small diameter tubes. Int J Heat Mass Transf 36:1269–1285
49. Hall DD, Mudawar I (2000) Critical heat flux (CHF) for water flow in tubes. Part II: subcooled CHF correlations. Int J Heat Mass Transf 43:2605–2640
50. Ong CL, Thome JR (2011) Macro-to-microchannel transition in two-phase flow: part 2—flow boiling heat transfer and critical heat flux. Exp Therm Fluid Sci 35:873–886
51. Zhang W, Hibiki T, Mishima K, Mi Y (2006) Correlation of critical heat flux for flow boiling of water in mini-channels. Int J Heat Mass Transf 49:1058–1072
52. Kosar A, Peles Y (2007) Critical heat flux of R-123 in silicon-based microchannels. J Heat Transf 129:844–851
53. Qi SL, Zhang P, Wang RZ, Xu LX (2007) Flow boiling of liquid nitrogen in microtubes: part II: heat transfer characteristics and critical heat flux. Int J HeatMass Transf 50:5017–5030
54. Revellin R, Mishima K, Thome JR (2009) Status of prediction methods for critical heat fluxes in mini and microchannels. Int J Heat Fluid Flow 30:983–992
55. Chung JN, Chen T, Maroo SC (2011) A review of recent progress on nano/micro-scale nucleate boiling fundamentals. Front Heat Mass Transf 2:023004
56. Boure JA, Bergles AE, Tong LS (1973) Review of two phase flow instability. Nucl Eng Des 25:165–192
57. Fukuyama Y, Hirata M (1982) Boiling heat transfer characteristics with high mass flux and disappearance of CHF following to DNB. In: Proceedings of the 7th international heat and mass transfer conference, vol 4, pp 273–278
58. Hosaka S, Hirata M, Kasagi N (1990) Forced convective subcooled boiling heat transfer and CHF in small diameter tubes. In: Proceedings of the 9th international heat and mass transfer conference, vol 2, pp 129–134
59. Jiang L, Wong M, Zohar Y (2000) Phase change in microchannel heat sink under forced convection boiling. In: Proceedings of the IEEE Micro Electro Mechanical Systems (MEMS), pp 397–402
60. Yen T, Kasagi N, Suzuki Y (2003) Forced convective boiling heat transfer in micro-tubes at low mass and heat fluxes. Int J Multiph Flow 29:1771–1792
61. Bowers MB, Mudawar I (1994) High flux boiling in low flow rate, low pressure drop mini-channel and micro-channel heat sinks. Int J Heat Mass Transf 37(2):321–332
62. Kew P, Cornwell K (1997) Correlation for prediction of boiling heat transfer in small diameter channel. J Therm Eng 17:705–715
63. Roday AP, Borca T, Jensen MK (2008) The critical heat flux condition with water in a uniformly heated microtube. J Heat Transf 130:012901
64. Koşar A, Kuo CJ, Peles Y (2006) Suppression of boiling flow oscillations in parallel microchannels by inlet restrictors. J Heat Transf 128:251–260
65. Koşar A, Peles Y (2007) Critical heat flux of R-123 in silicon-based microchannels. J Heat Transf 129(7):844–851
66. Kuan WK, Kandlikar SG (2006) Critical heat flux measurement and model for refrigerant-123 under stabilized flow conditions in microchannels. In: Proceedings of IMECE, ASME international mechanical engineering Congress and exposition, Chicago, Illinois, USA, IMECE-13310, 5–10 November
67. Rostami AA, Hassan AY, Chia SL (2000) Conjugate heat transfer in microchannels. In: Heat transfer and transport phenomena in microsystems, Banff, Alberta, Canada, pp 121–128
68. Celata GP, Cumo M, Mariani A (1997) Geometrical effects on the subcooled flow boiling critical heat flux. Rev Gen Therm 36:807–814
69. Nariai H, Inasaka F, Uehara K (1988) Critical heat flux in narrow tubes with uniform heating. Trans Jpn Soc Mech Eng 54(502):1406–1410

70. Bergles AE, Rohsenow WM (1962) Forced convection surface-boiling heat transfer and burnout in tubes of small diameter. Contract AF 19 (604)-7344 report, Department of Mechanical Engineering, Massachusetts Institute of Technology

71. Roach GM Jr, Abdel-Khalik SI, Ghiaasiaan SM, Dowling MF, Jeter SM (1999) Low flow critical heat flux in heated microchannels. Nucl Sci Eng 13:411–425

72. Oh CH, Englert SB (1993) Critical heat flux for low flow boiling in vertical uniformly heated thin rectangular channels. Int J Heat Mass Transf 36(2):325–335

73. Lazarek GM, Black SH (1982) Evaporative heat transfer pressure drop and critical heat flux in a small vertical tube with R113. Int J Heat Mass Transf 25(7):945–960

74. Yu W, Wambsganss MW, Hull JR, France DM (2001) Critical heat flux and boiling heat transfer to water in a 3mm diameter horizontal tube. In: Proceedings of the 2001 vehicle thermal management systems conference, paper no. 2001-01-1768

75. Lezzi AM, Niro A, Beretta GP (1994) Experimental data on CHF for forced convection water boiling in long horizontal capillary tubes. In: Proceedings of the 10th international heat transfer conference, Rugby, vol 7, pp 491–496

76. Harirchian T, Garimella SV (2009) The critical role of channel dimension, heat flux, and mass flux on flow boiling regimes in microchannel. Int J Multiph Flow 35:349–362

77. Harirchian T, Garimella SV (2009) The critical role of channel cross-sectional area in microchannel flow boiling heat transfer. Int J Multiph Flow 35:904–913

78. Kandlikar SG (2009) A scale analysis based theoretical force balance model for critical heat flux (CHF) during saturated flow boiling in microchannels and minichannels. In: Proceedings of ASME 2009 second micro/nanoscale heat and mass transfer international conference, Shanghai, China

79. Jacobi AM, Thome JR (2002) Heat transfer model for evaporation of elongated bubble flows in microchannels. J Heat Transf 124(6):1131–1136

80. Kandlikar SG (2010) Similarities and differences between flow boiling in microchannels and pool boiling. Heat Transf Eng 31(3):159–167

81. Revellin R, Thome JR (2008) A theoretical model for the prediction of the critical heat flux in heated microchannels. Int J Heat Mass Transf 51:1216–1225

82. Revellin R, Thome JR, Bejan A, Bonjour J (2009) Constructal tree—shaped microchannel networks for maximizing the saturated critical heat flux. Int J Therm Sci 48:342–352

83. Kosar A (2009) A model to predict saturated critical heat flux in minichannels and microchannels. Int J Therm Sci 48:261–270

84. Kuan WK, Kandlikar SG (2008) Experimental study and model on critical heat flux of refrigerants-123 and water in microchannels. J Heat Transf 130(3):1–5, 034503

85. Yen TH et al (2006) Visualization of convective boiling heat transfer in single microchannels with different shaped cross-sections. Int J Heat Mass Transf 49(21–22):3884–3894

86. Agostini B et al (2008) High heat flux flow boiling in silicon multi-microchannels—part III: saturated critical heat flux of R236fa and two-phase pressure drops. Int J Heat Mass Transf 51(21–22):5426–5442

87. Agostini B et al (2008) High heat flux flow boiling in silicon multi-microchannels—part I: heat transfer characteristics of refrigerant R236fa. Int J Heat MassTransf 51(21–22):5400–5414

88. Agostini B et al (2008) High heat flux flow boiling in silicon multi-microchannels—part II: heat transfer characteristics of refrigerant R245fa. Int J Heat MassTransf 51(21–22):5415–5425

89. Lee PC, Pan C (2008) On the eruptive boiling in silicon-based microchannels. Int J Heat Mass Transf 51(19–20):4841–4849

90. Bertsch SS, Groll EA, Garimella SV (2008) Refrigerant flow boiling heat transfer in parallel microchannels as a function of local vapor quality. Int J Heat Mass Transf 51(19–20):4775–4787

91. Lee PS, Garimella SV (2008) Saturated flow boiling heat transfer and pressure drop in silicon microchannel arrays. Int J Heat Mass Transf 51(3–4):789–806

92. Wang G, Cheng P (2009) Subcooled flow boiling and microbubble emission boiling phenomena in a partially heated microchannel. Int J Heat Mass Transf 52(1–2):79–91

93. Geisler KJL, Bar-Cohen A (2009) Confinement effects on nucleate boiling and critical heat flux in buoyancy-driven microchannels. Int J Heat Mass Transf 52(11–12):2427–2436

94. Yang ZL, Palm B, Sehgal BR (2002) Numerical simulation of bubbly two-phase flow in a narrow channel. Int J Heat Mass Transf 45(3):631–639
95. Mukherjee S, Mudawar I (2003) Smart pumpless loop for micro-channel electronic cooling using flat and enhanced surfaces. IEEE Trans Compon Pack Technol 26(1):99–109
96. Dupont V, Thome JR, Jacobi AM (2004) Heat transfer model for evaporation in microchannels, part II: comparison with the database. Int J Heat Mass Transf 47(14–16):3387–3401
97. Steinke ME, Kandlikar SG (2004) An experimental investigation of flow boiling characteristics of water in parallel microchannels. J Heat Transf 126(4):518–526
98. Kandlikar SG et al (2006) Stabilization of flow boiling in microchannels using pressure drop elements and fabricated nucleation sites. J Heat Transf 128(4):389–396
99. Kosar A, Kuo CJ, Peles Y (2006) Suppression of boiling flow oscillations in parallel microchannels by inlet restrictors. J Heat Transf 128(3):251–260
100. Kuo CJ, Peles Y (2008) Flow boiling instabilities in microchannels and means for mitigation by reentrant cavities. J Heat Transf 130(7):072402–072410
101. Kuo CJ, Peles Y (2009) Pressure effects on flow boiling instabilities in parallel microchannels. Int J Heat Mass Transf 52(1–2):271–280
102. Mukherjee A, Kandlikar SG (2009) The effect of inlet constriction on bubble growth during flow boiling in microchannels. Int J Heat Mass Transf 52(21–22):5204–5212
103. Zhang T et al (2010) Analysis and active control of pressure-drop flow instabilities in boiling microchannel systems. Int J Heat Mass Transf 53(11–12):2347–2360
104. Ajaev VS, Homsy GM (2001) Three-dimensional steady vapor bubbles in rectangular microchannels. J Colloid Interf Sci 244(1):180–189
105. Mukherjee A, Dhir VK (2004) Study of lateral merger of vapor bubbles during nucleate pool boiling. J Heat Transf 126(6):1023–1039
106. Mukherjee A, Kandlikar SG (2005) Numerical simulation of growth of a vapor bubble during flow boiling of water in a microchannel. J Microfluid Nanofluid 1(2):137–145
107. Lee W, Son G (2008) Bubble dynamics and heat transfer during nucleate boiling in a microchannel. Numer Heat Transfer, Part A 53(10):1074–1090
108. Suh Y, Lee W, Son G (2008) Bubble dynamics, flow, and heat transfer during flow boiling in parallel microchannels. Numer Heat Transfer, Part A 54(4):390–405
109. Kandlikar SG (2004) Heat transfer mechanisms during flow boiling in microchannels. J Heat Transf 126:8–16
110. Thome JR, Dupont V, Jacobi AM (2004) Heat transfer model for evaporation in microchannels, part I: presentation of the model. Int J Heat Mass Transf 47:3375–3385
111. Wang G, Cheng P, Bergles AE (2008) Effects of inlet/outlet configurations on flow boiling instability in parallel microchannels. Int J Heat Mass Transf 51:2267–2281
112. Harirchian T, Garimella SV (2010) A comprehensive flow regime map for microchannel flow boiling with quantitative transition criteria. Int J Heat Mass Transf 53:2694–2702
113. Lee PC, Tseng FG, Pan C (2004) Bubble dynamics in microchannels, part I: single microchannel. Int J Heat Mass Transf 47:5575–5589
114. Koşar A, Kuo C-J, Peles Y (2005) Reduced pressure boiling heat transfer in rectangular microchannels with interconnected reentrant cavities. J Heat Transf 127:1106–1114
115. Roday AP (2007) Study of the critical heat flux condition in microtubes. PhD thesis, Rensselaer Polytechnic Institute, Troy
116. Hasaan I, Vaillancourt M, Pehlivan K (2005) Two phase flow regime transitions in microchannels, a comparative experimental study. Microscale Thermophys Eng 9:165–182
117. Revellin R, Thome JR (2007) A new type of diabatic flow pattern map for boiling heat transfer in microchannels. J Micromech Microeng 17:788–796
118. Roday AP, Jensen MK (2007) Experimental investigation of the CHF condition during flow boiling of water in microtubes, paper no. HT2007-32837. In: ASME-JSME thermal engineering summer heat transfer conference, Vancouver
119. Katto Y, Yokoya S (1984) Critical heat flux of liquid helium (I) in forced convection boiling. Int J Multiph Flow 10:401–403

120. Lin S, Kew PA, Cornwell K (2001) Flow boiling of refrigerant R141b in small tubes. Trans IChemE, Part A 79:417–424
121. Pettersen J (2004) Flow vaporization of CO2 in microchannel tubes. Exp Therm Fluid Sci 28:111–121
122. Thorsen T, Maerkl SJ, Quake SR (2002) Microfluidic large-scale integration. Science 298:580–584
123. Kandlikar SG, Grande WJ (2004) Evaluation of single phase flow in microchannels for high flux chip cooling-thermohydraulic performance enhancement and fabrication technology. In: Proceedings of the 2nd international conference on microchannels and minichannels, ASME, Rochester
124. Kandlikar SG (2006) Effect of liquid–vapor phase distribution on the heat transfer mechanisms during flow boiling in minichannels and microchannels. Heat Transfer Eng 27(1):4–13
125. Daleas RS, Bergles AE (1965) Effects of upstream compressibility on subcooled critical heat flux, paper 65-HT-67, ASME, New York
126. Vafaei S, Wen D (2010) Critical heat flux of subcooled flow boiling of alumina nanofluids in a horizontal microchannel. J Heat Transf 132(102404):1–7
127. Liao J, Mei R, Klausner JF (2004) The influence of the bulk liquid thermal boundary layer on saturated nucleate boiling. Int J Heat Fluid Flow 25(2):196–208
128. Tomar G, Biswas G, Sharma A, Agrawal A (2005) Numerical simulation of bubble growth in film boiling using a coupled level-set and volume-of-fluid method. Phys Fluids 17(11):103–115
129. Genske P, Stephan K (2006) Numerical simulation of heat transfer during growth of single vapor bubbles in nucleate boiling. Int J Therm Sci 45(3):299–309
130. Son G, Dhir VK (2008) Numerical simulation of nucleate boiling on a horizontal surface at high heat fluxes. Int J Heat Mass Transf 51(9–10):2566–2582
131. Wu JF, Dhir VK (2010) Numerical simulations of the dynamics and heat transfer associated with a single bubble in subcooled pool boiling. J Heat Transf 132(11):501–515
132. Banerjee D (2009) Flow boiling in microchannels prepared as a part of two-phase flows and heat transfer term project. Presented to Texas A&M University MEEN 624
133. Velichala A, Vijaykumar A, Eniket E, Rajarova N Flow boiling in microchannels, IIT Kanpur, India
134. Talukdar P Boiling and condensation, IIT Delhi
135. Callao CM (2010) Flow boiling heat transfer in single vertical channels of small diameter. Doctoral thesis, Division of Applied Thermodynamics and Refrigeration, Department of Energy Technology, Royal Institute of Technology, Stockholm, Sweden
136. Kandlikar SG (2009) Similarities and differences between flow boiling in microchannels and pool boiling. In: Proceedings of the second micro and nano flows conference, West London, keynote contribution
137. Kadam ST, Kumar R (2014) Twenty first century cooling solution: microchannel heat sinks. Int J Therm Sci 85:73–92
138. Dhir V, Kabarajith HS, Ding L (2007) Bubble dynamics and heat transfer during pool and flow boiling. Heat Transf Eng 28(7):608–624
139. Roday AP, Jensen MK (2009) A review of the critical heat flux condition in mini-and microchannels. J Mech Sci Technol 23:2529–2547
140. Steinke ME, Kandlikar SG (2003) Flow boiling and pressure drop in parallel microchannels. In: Proceedings of first international conference on microchannels and minichannels, Rochester, New York, 24–25 April, pp 567–579
141. Boyd RD (1985) Subcooled flow boiling critical heat flux and its application to fusion energy components—part 1: a review of fundamentals of CHF and related data base. Fusion Technol 7:7–30

Chapter 3
Conclusion and Further Research

Knowledge about CHF is very important to deal with high heat flux for nuclear engineering and electronic equipment. Larger diameter and length of the channels are used for nuclear engineering because of large mass inventory. However, in recent times, the microchannels have been developed and the mass inventory inside the channels is usually very small. Single channel is seldom used, whereas multiple numbers of parallel channels are more often used. CHF increases with mass flux, channel hydraulic diameter for saturated boiling (and inversely for subcooled boiling), length-to-diameter ratio of the channel, saturation temperature and degree of subcooling at channel inlet. Modern machining techniques are now-a-days being routinely used for the fabrication of microchannels. Research with microchannels are using high speed photography with relatively high magnification to capture the details of flow regimes, bubble growth and measuring liquid film thickness. However, widely applicable correlations are needed to predict data accurately. The microchannel heat transport phenomena are still far from being clearly understood. More accurate measuring techniques are necessary. More and more finer models are necessary to predict CHF accurately.

© Springer International Publishing Switzerland 2015
S.K. Saha, G.P. Celata, *Critical Heat Flux in Flow Boiling in Microchannels*,
SpringerBriefs in Applied Sciences and Technology,
DOI 10.1007/978-3-319-17735-9_3